A LEGACY OF NAMES

BRITISH PLACES IN THE NEW WORLD

Elaine Borish, an American living in London, was born in New York City. She holds degrees from Rutgers, Boston University, and Northeastern University and has taught at universities in New England. In old England, she has lectured in English and American literature at Morley College in London. Her numerous articles have appeared in leading newspapers and magazines.

Also by Elaine Borish

Literary Lodgings
Unpublishable!

Elaine Borish

A LEGACY OF NAMES

BRITISH PLACES IN THE NEW WORLD

Fidelio Press

Boulder • London

Published by
Fidelio Press
61 Pine Tree Lane
Boulder, Colorado 80304

ISBN 0-9524881-2-4

Printed in the United States of America.

Cover design by Allee Trendall

Illustrations by Sheila Acton

Contents

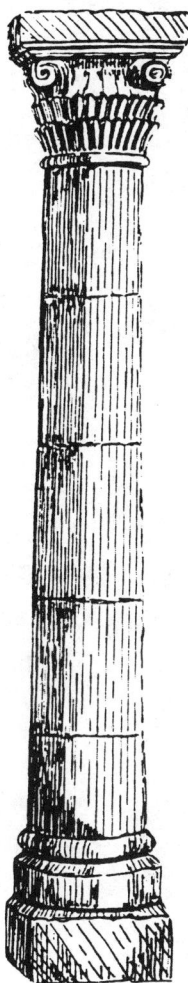

To the memory of
Phyllis
who loved travel

Preface

What's in the name of a town, the poet might have asked. A great deal—is the answer I might offer. This book is for anyone whose curiosity or imagination is caught by the sources and associations of similar place names.

Many American cities and towns were named for the places in England from which the first settlers emigrated. Early citizens took their familiar homeland names with them to the new world and established a permanent bond with the native town left behind. Americans still feel the alliance deeply.

That Americans remain intrigued with the familiar place names in Old England can be witnessed by the steady streams of modern pilgrims who flow into such places as Plymouth or Groton, Braintree or Boston—in England. Regrettably, the number of visitors to other enjoyable towns with such evocative names as Framlingham, Shrewsbury, or King's Lynn are not so great, but perhaps the accounts given in this book will entice readers and increase the influx of visitors to less touristy places. However, even the armchair traveler can enjoy reading about the linking of one place with another. How did the name come to be chosen? What is the English town like today?

This is not a definitive study. Nor is it possible to list with certainty all of the connections that do exist. The selections, merely representative, give the meaningful associations of the English counterpart and a feeling of the English town as it now exists.

Andover

Once a quiet country market town in northwest Hampshire, Andover was greatly expanded in the 1960s when it became a center for overspill from London, only an hour's train ride away and sixty-five miles by car. Population increased, industries grew, housing estates proliferated, and shopping centers spread. A maze of roads now encircles the town where housing and industrial complexes have sprouted; but, fortunately, the old center has not been destroyed.

An attempt was made to preserve the town's former shape and character as well as the dominance of its two focal buildings, the Guildhall and the Church of St. Mary. The classical Guildhall of 1825 survives in the market place at the head of the wide High Street, while the narrow Upper High Street leads to the parish church that overlooks the old town from its position on the slight hill.

Perhaps the best Victorian church in Hampshire, St. Mary's was completed in 1846 in Early English style with an impressive interior of lofty arches, plaster vaulting, and chancel seen through an elaborate screen.

Among the Georgian and older buildings surrounding the church is the Andover Museum, which features exhibits on local history and geology and houses also the Museum of the Iron Age. It displays artifacts from the hill fort known as Danebury Hill, located just a few miles from the town center. Reconstructions of the ramparts, a round house, storage pits, and a grave are aspects of the life of the prehistory encampment that occupied the site over two thousand years ago. (Surely, it was Andover's Iron Age hill fort that gave its name to Danbury, Connecticut.)

Andover, a Celtic word meaning spring water, derives from the River Anton that flows through the town. A prehistoric highway crossed the river at the Andover site before settlement took place. It is tempting, if fanciful, to visualize traders and

1

primitive travelers on their way to Stonehenge over the track known as Harringway. In later centuries, that major trackway, the reason for the initial development of Andover, was the route for pilgrims on their way to Canterbury.

A church existed at Andover in 994. Henry II granted a charter in 1178, and the town grew steadily. During the Middle Ages, Andover was the third largest in Hampshire, exceeded only by Winchester and Southampton. The medieval town grew around a market near the church and priory, and the wide High Street served as a later market site. The Saturday market, granted by a charter of 1599, still takes place below the market clock.

During the coaching era, because Andover was a natural staging post on the London to Exeter road, the town expanded even more rapidly. Formerly known as the Star and Garter, the Danebury Hotel dates from the same time as the nearby Guildhall. An attractive inn with Tuscan porch and bow windows, the Danebury is where George III stayed the night on his annual journey to the seaside. Original frontages survive in this part of the High Street, which culminates in St. Mary's on the hill.

The George, the Globe, and the Angel are among several older hostelries with yards that are integral to coaching establishments. Gone are the horses; but the Angel, rebuilt in 1445 when it was known as College Inn, claims a place in history for having served a meal to James II in 1688. While old inns were preparing dinners, royal or otherwise, Andover was being formed in America.

At least fifteen Andovers exist in America, but it is the one in Massachusetts, which came into being on 6 May 1646 by act of the General Court, that was the first to be called Andover. The name was selected by settlers from Andover in England to replace the Indian designation of Cochichawicka (or Cochichewick), and the name was subsequently applied to other American settlements. Connecticut's Andover was also named after the English mother town in the county of Hampshire, as perhaps were one or two others. But the more frequent pattern was for migrating residents to take the name of Andover with them.

For example, in New Hampshire, New Breton had existed from 1751 until a new charter of 1779 changed it to the Town of Andover. Petitioners had arrived from southern New Hampshire near the Massachusetts border, making that state's Andover suspect as the source for the name.

Andover in New York may have been named for the one in Vermont after Thaddeus Baker, a resident who sold his holdings in Pultny, moved with his family in 1807. Other Vermonters joined the new Andover settlement in New York.

The Township of Andover, Maine, was purchased in 1791 from Massachusetts by Samuel Johnson and others from Andover, Massachusetts.

A woolen mill in Massachusetts gave North Andover, Wisconsin, its name in 1867. Virginia named its Andover for Andover College in New England. Illinois named it for the Massachusetts town. South Dakota, the westernmost of all, was named for the railroad superintendent's New England home.

So the former English backwater with a popular name has become a popular place for retracing heritage routes. Mayors and Rotary Clubs have exchanged visits; and, although Andover does not have the appeal of other Hampshire towns such as Winchester, Beaulieu, or Lymington, tourists searching for roots and making the relatively short journey to Andover from London find a great number of appealing things to behold in the traffic-free center, which still has the feeling of an old market town.

Barnstaple

Never mind the tasteless traffic patterns and tall buildings that characterize modern Barnstaple. Never mind the shops, department stores, and office buildings that line the High Street. Or the bus station built in the once bustling quay. It is still possible to visualize the active harbor scene of an earlier era and to become immersed in the economic activity and wealth that once characterized the old town.

With over a thousand years of history, Barnstaple, the North Devon town on the River Taw, is one of the oldest towns in Britain. In 930, it was an established Saxon borough known as Beardestaple, the Staple—Boundary or Market—of Bearda. It prospered over the centuries as an important market town and seaport, particularly for the expanding wool- and cloth-making industry, which made the quay an especially active place and the center of town life. Today, Barnstaple has taken on the predictable characteristics of ordinary, busy city life.

Evidence of the rich past is present everywhere. Spacious and elegant houses can be seen in the area around Trafalgar Lawn. High Street too has a number of later but interesting buildings including the Guildhall, built in 1826 in classical style. The School of Art at 42 High Street is an early nineteenth-century house. For contrast, the Three Tuns Tavern goes back to the fifteenth century. The Westminster Bank, with outstandingly fine plaster ceilings, was the House of Spanish Merchants much earlier in its career. And on Tuesdays and Fridays the seventeenth-century Pannier Market is worth a visit. The vast cast-iron and glass-roofed area is filled with stalls where farmers' wives sell a variety of local products—Devon cream, cheese, preserves, fruit, vegetables. Just opposite the market is Butcher's Row, an arcade of open-fronted butchers' shops built in mid-nineteenth century but retaining the medieval practice of grouping shops of one kind together.

4

The thirteenth-century Church of St. Peter, in the center of town, contains some unusual monuments which attest to the wealth of the seventeenth-century mercantile families. Elizabeth Delbridge is represented as she died in 1628 with her swaddled infant "of which she died in childbirth." Another monument on the wall between south aisle and chapel depicts a man comfortably in bed with his wife and child.

Next to the church is the fourteenth-century Chapel of St. Anne, now functioning as a small local museum. After it was suppressed by Henry VIII in 1547, the chapel was used as the town's grammar school for nearly four hundred years and is the place where John Gay received his education.

Born in Barnstaple in 1685, John Gay is known primarily as the author of *The Beggar's Opera*. Barnstaple is understandably proud of its local literary celebrity and announces its claim, even in such mundane ways as the "John Gay Coffee Shop."

The appropriately-named Long Bridge of thirteen arches spans the River Taw and the ages. Dating from the thirteenth century, and rebuilt around 1417, it has been widened and improved and altered on a number of occasions through the centuries, but it manages to maintain much of its medieval work as well as its picturesque qualities.

The Long Bridge

5

The river bank between Long Bridge and the quay constituted the old and lively town center with its pleasant colonnade known as Queen Anne's Walk. In front of the walk is the quay from which "six ships from North Devon joined Sir Francis Drake's fleet at Plymouth and helped to defeat the Spanish Armada," proclaims a plaque on the wall. Built as a merchants' exchange in 1609, the graceful walk was restored in 1714 with the addition of a statue of the good queen. Under Queen Anne's statue is the Tome Stone. Again, a plaque explains its use. Elizabethan merchants sealed bargains by placing payment on the mushroom-shaped stone before witnesses.

Queen Anne's Walk

Seven miles northeast of the town is Arlington Court, an elegant Regency house notable for its collections of shells, model ships, pewter, and carriages. The ancestral home of Sir Francis Chichester, famous navigator knighted for his epic voyage around the world in 1966, is now open to the public as a National Trust property. But perhaps the event of most interest to Americans is a particular seventeenth-century event, the founding of Barnstable, Massachusetts. In 1639, the General Court established three Cape Cod settlements, replacing their original Indian names with those of English ports: Barnstable,

6

Sandwich, Yarmouth. At the time when Barnstable was being settled, Barnstaple was enjoying the great prosperity which can still be seen in its elegant old houses.

The twin cities, Barnstable and Barnstaple, share not only the same names, but also similar appearances and functions. Both are set in harbors that look remarkably alike. Both are gateways—to Cape Cod and to North Devon—and centers for summer holiday activities. When Barnstable celebrated in 1939 the tercentenary of its founding, old Barnstaple presented its younger counterpart with a replica of one of its treasures, an exquisite silver-gilt Steeple Cup made in 1589 during the reign of Queen Elizabeth I. All of this may well conjure up in America thoughts of that other Town of a Thousand Years.

Bedford

The easy-to-reach county town of Bedford is located fifty miles north of London in Bedfordshire, a small county which even the guide books admit lacks the rich scenery so characteristic of much of England. "The least picturesque of the English counties," says the Blue Guide to England. Perhaps Bedford is spoiled by the approach to it. Factories and beds of clay, the basis of a large brick industry, mar and scar the landscape. With such bad press notices and with such an unappealing preview, one may wonder, why bother to reach it at all?

Because Bedford is nevertheless a pleasant town on the pretty, meandering River Ouse, with an interesting, colorful, and ancient background. Pepys recorded his visit to the town in 1667 and called it "a good country town." It still is.

Bedford has made excellent use of its tree-lined river, and the locals are justifiably proud of the Embankment with its riverside walks and gardens on both sides and seats which provide restful oases.

The five-arched bridge over the Ouse enhances the total picture. Turner painted Bedford Bridge, as have many others. The attractive scene takes in the Swan Hotel, built for the Duke of Bedford in 1794. Farther along the Embankment, just beyond the Swan, is the Bedford Museum with its specimens of local history. Nearby, the Cecil Higgins Art Gallery has a surprising collection which includes works by Turner and Dürer, bronzes by Epstein and Henry Moore. Plenty to do and see in Bedford without even leaving the river area!

A river crossing undoubtedly existed from the earliest times. The ancient town of Bedford was inhabited by Celtic and Roman settlers before being occupied by the Saxons in the seventh and eighth centuries. The river remained fordable here, and the Danes made frequent incursions on Bedford. They devastated the town in several raids, and the destruction by fire in 1010 is still in evidence on the tower of St. Peter's Church which bears

8

the marks of that fire. Shortly after the Norman Conquest, a castle with a commanding view of the river was built, of which only the mound remains. Old Bedford Bridge, which superseded a timber bridge, was built in 1224 from stones of the demolished castle. The present bridge dates from 1813.

Bedford Bridge

But the episode in history that dominates the Bedford of today centers on John Bunyan. Bunyan was born at Harrowden, just south of Bedford. The son of a tinker, he himself became a tinker and a thinker who developed deep religious convictions. In the village of Elstow, he spent his childhood, and there he remained after marriage until he moved a mile away, to Bedford, in 1655.

Elstow itself is worth the journey for its thirteenth-century abbey church, thirteenth-century cottages, and picturesque sixteenth-century Moot Hall. The ground floor of the Moot Hall contained six small shops to be used at fair times, while upstairs

Elstow Moot Hall

was the village meeting hall. Now the remarkable building houses a museum of seventeenth-century life associated with the time of John Bunyan.

Bunyan began preaching in 1656 and continued to do so even after nonconformist preaching was made illegal. A plaque at the corner of Silver Street in Bedford indicates the site of the County Gaol where he was imprisoned from 1660 to 1672 for religious dissent. Another plaque on the bridge sites the Town Gaol where he spent a further brief period of imprisonment in 1677 and where he probably began work on his best seller, *The Pilgrim's Progress*.

Remembrances of the man are everywhere. The Bunyan Museum adjoins the church. The Bunyan Meeting House in Mill Street was rebuilt in 1850 on the site of the earlier meeting place that he became pastor of upon release from prison in 1672. Its bronze doors, which recall the Baptistry doors in Florence, illustrate in bas-relief ten scenes from *The Pilgrim's Progress*. Bunyan's chair and the pulpit from which he preached are here, as is the door of his cell in the County Gaol.

Since Bedford achieved a claim to fame by imprisoning John Bunyan, it seems a form of justice that another Bedford worthy, John Howard (1726-90), became a great prison reformer. His statue, sculptured in 1894 by Sir Alfred Gilbert, the creator also of Eros in London's Piccadilly Circus, stands in St. Paul's Square.

Here in the town center of St. Paul's Square, just over the bridge and around the bustling market square, are Shire Hall, Town Hall, Corn Exchange, and—of course—St. Paul's Church.

St. Paul's is the major church of Bedford. With its embattled clerestory and aisles, it makes a handsome background for the statue. In the south chapel is a brass to Sir William Harpur, who died in 1573, and his wife. Harpur amassed a fortune in London, became Lord Mayor in 1561, and left a bequest for the establishment of almshouses and other charities. The Harpur Trust now supports four schools in Bedford. The facade of the Bedford Modern School in Harpur Street, built in 1833, has been retained for a large and handsome, newly-developed shopping complex.

In contrast to the over-restored St. Paul's, the church of St. Peter De Merton has Anglo-Saxon traces and a Norman central tower. On St. Peter's Green is the dominating motif of Bedford, a statue—as you might have guessed—of John Bunyan.

Bunyan lived in a time of religious persecution when many dissenters left from Bedford as well as from surrounding villages. Peter Bulkeley, Puritan rector of Odell, ten miles northwest of Bedford, emigrated to Massachusetts in 1635. From Cambridge, he led a party of planters through the woods to found a new colony which was named "Concord."

That the New World beckoned Bedfordshire people would seem to be indicated by the fact that Bedford is the name of a number of American cities—in New Jersey, Pennsylvania, Kentucky, Iowa, Indiana, Montana, Illinois, as well as Massachusetts.

Beverley

The "beaver clearing in the woods" has undergone thirteen hundred years of history to exist today as one of the prettiest market towns anywhere. Sir John Betjeman, the late poet laureate, enthusiastically summed it up as a "great surprise" for visitors and "a place made for walking in and living in."

Government reorganization in 1974 abolished the East Riding county, so that Beverley properly belongs now to the county of Humberside. But the three parts of Yorkshire—the Danish *Thridings*—were created long before the Norman Conquest, and old ways persist. Yorkshire people continue to cling to the old boundary names of North Riding, West Riding, and East Riding.

Beverley was the leading town of the East Riding in the Middle Ages. Mentioned in the Domesday survey, the small town began to prosper by the twelfth century from trade in Yorkshire wool and from cloth made in Beverley itself.

When wool ceased to play a major part in the life of the town, Beverley declined and was eventually overshadowed by Hull. But signs of former wealth of the Middle Ages are still visible. The Woolpack Inn in Westwood Road is a reminder of those prosperous days, as are some street names such as Walkergate (where cloth was washed by being walked through water), Flemingate (which recalls continental trade), and Dyer Lane.

Of the five medieval gates which once existed, the picturesque North Bar, rebuilt in 1409, survives. The ancient gateway leads into North Bar Within and to the fourteenth-century St. Mary's, the sole survivor of the town's three medieval parish churches. North Bar Within, the main thoroughfare, wends its way through the town from St. Mary's at the north end, past the eighteenth-century Market Cross at the center, to the venerable old Minster at the south end. It leads past the Saturday Market, narrow lanes, and gracious Georgian and medieval houses that justify Beverley's reputation as one of England's loveliest towns.

The present Minster was built on the site of a monastery founded by John, Bishop of York, who sought secluded retirement in Beverley in the year 719. The miracles he performed in his lifetime, recorded in the Venerable Bede's *Ecclesiastical History of the English People*, brought a great influx of pilgrims to worship at his shrine and led to his canonization in 1037. The original wooden monastery, sacked and destroyed by Viking invaders, was refounded as a minster church to serve the district.

Beverley Minster as seen today, richly adorned with statues on its fifteenth-century west front, dates from the thirteenth and fourteenth centuries. The two towers of the Minster dominate Beverley from any direction. Among its interior features are the fourteenth-century Percy Tomb and the misericords, or seats of mercy, exhibiting some magnificent wood carvings.

Only a short distance from the Minster's fine east window, tucked away and almost out of sight, is the medieval Dominican Friary, now a youth hostel.

Elsewhere in the town, at Number 25 Highgate, lived the deLancey family, prominent society people who emigrated to New York City and whose name is preserved in the famous Delancey Street of New York's lower east side. Of greater fame in the annals of name transference, is the adoption of the name of the town itself.

In America, the original Indian territory of Naumkeag included Beverly, which was incorporated as a distinct township in October 1668. Townspeople were apparently dissatisfied with the name, for Roger Conant petitioned the court in 1671 to change it to Budleigh, the Devonshire town in which he was born. He pleaded, furthermore, that the unfortunate name of Beverly was often distorted and derisively called by the nickname, *Beggarly*. The request was refused, and the name that derived from that Beverley of considerable note in the East Riding of Yorkshire, happily prevailed.

Billericay

A visitor to Billericay, that old-world Essex town just twenty-five miles northeast of London, need not search very long for a feeling of involvement in American history, for in 1620 a group of residents left their homes in Billericay's High Street to board the *Mayflower* for the New World.

It was Christopher Martin of Billericay, one of a contingent of four or five of the city's natives, who financed the *Mayflower* adventure and provisioned the ship. The group included Martin's wife, the widow Marie Prower whom he had married in 1607; his stepson Solomon Prower; John Langerman, a bondservant who worked for a fixed period of time in exchange for payment of expenses; and possibly a Peter Browne.

Christopher Martin's signature was ninth on the list of elders witnessing the *Mayflower* Compact written before the ship landed to found a Christian settlement in the wilderness. Alas, he was among the many pilgrims who succumbed during the first dreadful winter. William Bradford recorded of him: "He and all his, dyed in the first infection not long after the arrival" including an infant son born on the voyage.

Among others of Billericay who followed to New England was a Ralph Hill, to whom can be attributed the founding and naming of Billerica, Massachusetts, in 1655. In fact, when townships were being established in Massachusetts by emigrants largely from the eastern counties of England, as many as three dozen were given East Anglia names.

If it is meaningful to determine what's-in-a-name, St. Mary Magdalen Church on the High Street may be the best place to begin. It is on the site of the Chantry Chapel founded in 1342 in order to accommodate the people of Billericay for whom the parish church located some two miles away was inaccessible. It also served the needs of pilgrims passing through on their way to Canterbury or to the Shrine of Our Lady of Walsingham. Lands given to the support of the Chantry and its priest included the manor of Ramsden Crays, from which the name of the town may be derived. Simon de Crais owned the manor in the thirteenth century when it was Vill de Crais. Myriad spellings, including Villdecrey and Billerecrais, hint at an evolution in the name to Billericay. It is to that conjecture that the American "Billerica" can be traced.

Entirely rebuilt in 1490, the chapel underwent considerable alterations over the years until it was largely rebuilt in Georgian style in the 1780s. As seen today, the interior is reminiscent of a little New England church, with balcony and characteristic simplicity of style and decor.

Further New England reminders appear in the nearby Chantry House with its early sixteenth-century paneling, its seventeenth-century fireplace and exposed beams (probably recycled ships' timbers), and its twentieth-century pub. Of significance to New World history, Christopher Martin is said to have lived here.

No English town has charm without its inns and pubs, and Billericay has its share of charm. The Chequers is an old world inn of the sixteenth century, of timber construction, with low ceilings and beams. Located opposite the Chantry House, it undoubtedly entertained and lodged visiting priests among its guests.

The Crown Hotel has been licensed for at least four hundred years, although it is now on its third site. Dick Turpin, that infamous highwayman born in Essex in 1705, is associated with Billericay, specifically with the Crown. Here he is supposed to have ridden his horse up the staircase, in anticipation of the American Western, and jumped out of an upstairs window to make his escape. Naturally, his ghost still rides Billericay roads.

Returning to more sober ideas, off Chapel Street is Mayflower Hall, built in 1920 by the Congregational Church to commemorate the tercentenary of the sailing of the Pilgrim Fathers. A plaque in the entrance establishes the part played by Billericay in the *Mayflower* venture. The Congregational Church itself is nearby.

But Billericay is more than just a source for Americana. Archeological evidence exists of man's presence here in the Stone Age. There is evidence too of a prehistoric hill-fort settlement. In a later era, the site was a Roman villa. Billericay had a market in the thirteenth century and two fairs, indicating that it was an important and thriving town in the Middle Ages. On the unspoiled main street are preserved a number of buildings of interest, a few dating to the fifteenth century. Fine Georgian houses are preserved as well—all attesting to a prosperous past. The Cater Museum contains a diversity of displays such as English coins of the twelfth century, rooms furnished in authentic Victorian fashion, and fragments of a German Zeppelin shot down in the area in 1916. Among the maps, sketches, data, and information illustrating the town's history is a photograph of Billerica in Massachusetts.

So it would seem that American associations are dominant. At one time, the Chantry House was an English restaurant serving such traditional specialties as Pork Mayflower and Petit Poussin Pilgrim, but it was later replaced by an Indian Restaurant—marking at least one case in which the Indians won out over the Pilgrims.

The Chantry House

Birmingham

There are those who believe that the best thing about Birmingham is the opportunity to escape from it into surrounding green countryside or to such inviting nearby places as Stratford or Warwick. They see the industrial city as a dirty and ugly place of cheap houses and pollution, traffic and spaghetti junctions. But there are also those who believe the city has been too much maligned. No doubt, Brummagem, as locals affectionately refer to their city, has many offerings for those who escape to it from London, 111 miles to the south. And certainly, with the recent refurbishment, loyal Brummy views are justified.

The rebuilding of Birmingham has transformed the city and made a Victorian theatrical song of 1828 seem eerily applicable today:

> Full twenty years and more, are past,
> Since I left Brummagem;
> But I set out for home at last
> To good old Brummagem.
> But every place is altered so,
> There's hardly a single place I know;
> And it fills my heart with grief and woe,
> For I can't find Brummagem.

In the old hub of Birmingham, a city famed for markets, the Bull Ring is the site of the original market established in 1166 when Henry II granted a charter to Peter de Bermingham for a market which has been thriving ever since. The name derives from an iron ring set in the ground near the row of butchers' shops, or shambles, to which the bull would have been tethered before being slaughtered.

The de Berminghams, former lords of the manor, lie buried in the Parish Church of St. Martin, which has stood on the ancient market site at least since its earliest recorded date of 1263. To those fourteenth-century lords goes the credit for supplying names of other people and places, over the centuries, and to American cities in particular.

The flourishing thirteenth-century market town was destined to become an industrial city. Even in Shakespeare's time, large numbers of smiths used coal from the mines of the neighboring county of North Warwickshire as fuel for hammering out their wares. When the Elizabethan historian Leland visited, he was most impressed with the noise of anvils and recorded it in a sixteenth-century account which might make the modern reader wonder whether the composer Verdi found in Birmingham the source for his famous Anvil Chorus. The textile and leather industries were also of considerable importance, and a further expansion came about when the Industrial Revolution made Birmingham one of the greatest manufacturing centers in the world and the second largest in Britain.

Among old buildings that have managed to survive despite industrialization are the stately Jacobean mansion of Aston Hall located in suburban Aston Park and the Cathedral of St. Philip. Built as a church in 1715, it became a cathedral in 1905; the English Baroque edifice, surrounded by greenery, contains Victorian stained glass by Sir Edward Burne-Jones (who was born in Birmingham) and made by William Morris.

Belonging to the Victorian city are the Great Western Arcade, the Victoria Law Courts, the northern end of Colmore Row, the southern end of Corporation Street, and St. Chad's Roman Catholic Cathedral of 1839, the first Catholic cathedral to be built in England after the Reformation.

Also Victorian is the Town Hall, based on the temple of Castor and Pollux in the Roman forum. Here in the Town Hall, Mendelssohn conducted his *Elijah* in 1846, written for a Triennial Festival; and Sibelius conducted his Fourth Symphony. Yehudi and Hephzibah Menuhin played here to thunderous ovations.

Today, the city is also the home of the well-known Birmingham Repertory Theatre, the Birmingham Royal Ballet, the D'Oyly Carte Opera Company, and the renowned City of Birmingham Symphony Orchestra. Founded in 1920 and made famous under the direction of Sir Simon Rattle, the orchestra plays in the new and acoustically superb Symphony Hall of the city known for good music.

The 2,200-seat Symphony Hall in the city center opened in 1991, a year of rebirth which also saw the launching of the £160 million International Convention Centre and a new National Indoor Arena for sports and entertainment. The city may well boast of unrivalled exhibition, event, conference and convention facilities.

At the heart of the revitalized city is the pedestrianized piazza known as Centenary Square. Embellished with a fountain and sculptures, the attractive traffic-free center has smart shops in a complex of malls called Paradise Forum, offices, cinema, restaurants, pubs, and hotels.

Also in the center, the Birmingham Museum and Art Gallery contains an outstanding collection in which are represented such masters as Botticelli, van Ruisdael, Renoir, and Gainsborough as well as a major collection of pre-Raphaelite paintings.

For those in search of further cultural or academic pursuits, there is the world-respected Birmingham University in the Edgbaston area. The Chamberlain clocktower, in Tuscan campanile style, soars in the center of a vast campus where the Barber Institute of Fine Arts exhibits an impressive collection of paintings—Turner, Rubens, van Dyck, Murillo, Monet, Degas. Also located in the handsome nineteenth-century suburb of Edgbaston is the Botanical Gardens, featuring tropical glasshouses.

In an area known as the Jewellery Quarter, a concentration of factories and workshops has been producing silverware and jewelry for well over a hundred years. Nowadays, well over a hundred shops offer an unusual shopping treat as they purvey enticing baubles and bangles of gold and silver. The historic district also displays working conditions and craftsmanship in a Discovery Centre.

A culminating bonus of the industrial heritage is the city's network of canals dating from the 1780s. These lovely waterways still provide possibilities for pleasant walks on renovated towpaths or cruises on traditional narrowboats.

The city has certainly come a long way from the Domesday survey of 1086 which valued the dwelling place or ham of the family of Beorma, with its population of thirty-two persons, at twenty shillings.

So let others deride Birmingham's industrial noise and pollution and twentieth-century traffic disasters. In the heart of England, Britain's second largest city has a myriad of attractions to delight visitors, not the least of which is the opportunity to feel a deep sense of satisfaction that comes from exploring away from summer throngs who overcrowd the customary tourist places. At the same time, the visitor may pay tribute not only to the twenty Birminghams that exist in the United States, but also to the Birmingham Crater on the moon.

Boston

Old Boston in England is, naturally, a popular pilgrimage place for Americans. Indeed, the stranger in Boston is soon made overtly aware of ties to the Massachusetts namesake. American accents may be heard everywhere, American addresses may be read with surprising frequency in various visitors' registers, and the American flag may even be seen flying on several buildings.

If the visit is undertaken as a pilgrimage to the past, to the source of new Boston, the tour could appropriately start in the old Guildhall. Inside the fifteenth-century Guildhall are the iron-gated cells which once imprisoned seven "offenders" including William Bradford and William Brewster. The group, later to be known as the Pilgrim Fathers, was betrayed by the captain of a Dutch ship and arrested as they attempted to sail in 1607 from Boston to Holland and thence to the New World. They were tried in the courtroom above the cells. A more successful attempt followed in 1630, and the settlement established was named "Boston" to commemorate the associations which were so meaningful to the Massachusetts colonists.

But the naming of old Boston in Lincolnshire goes back to the year 654 when a considerate and popular Benedictine monk, St. Botolph, requested a site in the uninhabited, desolate fenland in order not to evict residents from lands they possessed. When his monastery was destroyed by the Danes in 870, it was rebuilt and the area called St. Botolph's Town. From that tribute to its founder, the name can be traced to the contracted and corrupted form in present usage.

Boston today, with a population over 26,000, is a prosperous country town situated in a highly productive agricultural area. Situated also on the River Witham, Boston is a busy port and a center for a thriving shell fish industry. Its location and surrounding fen country, characterized by stretches of watery fields, recall the American Boston, where a certain area in the marshy Back Bay came to be called the Fenway. It was largely in the eighteenth century that huge areas of the fens were drained by canals and dykes and man-made watercourses and enclosed for farming purposes. The centrally located Market Place, with its stalls selling fresh produce, serves as a perennial reminder of agricultural yield. As an extension of crop production, the canning of fruits and vegetables is a major Boston industry.

But this is not a modern traffic-clogged city, for there in the center of downtown Boston, in Willoughby Street, is the Maud Foster Windmill. Unusual for its five sails or sweeps, it was built in 1809 to grind corn. It is still in working order and remains a fine example of just one of the many mills which once abounded.

Among many other Dutch associations and influences is a thriving tulip industry. Over 10,000 acres of tulip fields in Lincolnshire rival those in Holland for production and beauty and receive a great influx of visitors in the springtime. One area of the county, to the southeast, is even called Holland for the reason that it too can be characterized as hollow or flat land.

But one need not leave the center of town to enjoy the pleasant impression of Boston itself, which is conveyed in the area around the irregularly-shaped Market Place. Architectural variety can be seen all around in such buildings as the Grand Peacock and Royal Hotel of about 1670, or the Exchange buildings dated 1772, or the Assembly Rooms of 1826 with Tuscan columns and tall windows lighting a large assembly room.

Colorful though the area may be on market days, the scene is dominated, as it has been for centuries, by the enormous St. Botolph's Church with its soaring lantern tower, affectionately known to all as the "Stump." This prodigious parish church is 282 feet long and 100 feet wide with a tower that rises to a height of 272 feet.

Why the exceedingly high tower is called the Stump no one knows. But the tradition of the name may be the best evidence that a spire was originally intended to top the whole. Perhaps it reflects modesty of Bostonians. Or perhaps it was the envious expression of neighbors, for a claustrophobic climb to the first balcony of the tower can give some remarkable views of the

Boston Stump

town and fenland—one third of the county—including, on a clear day, Lincoln Cathedral, some thirty miles off. The ascender can look down at the red roofs and confused maze of streets or follow the course of the Witham or look northwards to the Lincolnshire of Tennyson's childhood.

In fact, the open stonework gives this medieval lantern tower a rather fragile appearance for its practical function. Designed to act as a guide to mariners out at sea and to travelers who would see it across the fens, it obviates any idea of originality attributed to the new Boston patriots who used a lantern in the Old North Church steeple as a warning guide: "One if by land and two if by sea. . . ."

Inside, the tower again makes an unforgettable impression as it opens up to a height of 137 feet. American associations pile up too. In the tower area, is a memorial to five Boston men who later became governors of Massachusetts. Ironically, the need for repairs helps to bind the two Bostons. In 1931, Americans donated a generous sum for restoration of the tower. And earlier, in 1857, the people of Boston, Massachusetts, were responsible for restoring one of the former guild chapels in memory of John Cotton, who was vicar of St. Botolph's from 1612 until he left for the other Boston in 1631.

Begun in 1309, St. Botolph's Church was completed nearly one hundred years later, only to have rebuilding start all over again. Renovations and changes, which have been going on continually, tie the two Bostons in both directions. Tracery from one chancel window in the east end was removed and shipped across the ocean to be installed in a cloister at Trinity Church in Copley Square, Boston.

The reason for the enormous size of St. Botolph's can be attributed to the equally enormous prosperity which the town achieved in the fourteenth century. Its position on the east coast established Boston as an important port as early as 1204, when King John granted its charter. Trade boomed and by the middle of the fourteenth century, Boston was second only to London as the busiest port in the country. Wool was the chief export. Hanseatic merchants became well established, and prosperous guilds paid for the building of the church. However, the wool trade did not survive. Boston began a sad decline first with a series of plagues and then in the early fifteenth century when floods caused the silting up of the river. As a direct result of the great flood of 1571, Boston became a distressed area. The town rallied again, particularly after the drainage of the northern fens, but the prosperity of the eighteenth century was only moderate in comparison. The disproportionately giant church remains as a symbol of former prosperity brought about by the wealthy wool trade. Significantly, the steeple faces the river, as if looking toward the port and source of Boston's wealth—the reason for its existence.

In the attractive and famous town of Boston, ruins also survive to summon up the past. Friars settled here, as they did in all thriving medieval towns, and the remains of Blackfriars are now to be found in Spain Lane, once part of the thirteenth-century Dominican friary and now converted into a theatre.

The Guildhall, of course, serves as a well-preserved and poignant reminder of the past. Just next to it is the Fydell House, built in 1726, predictably by William Fydell, a successful wine merchant who was also three times mayor of Boston. It is a superb town house and an excellent example of eighteenth-century domestic architecture. One room, opened in 1938 by the American ambassador, the Hon. Joseph P. Kennedy, has been designated for the use of American visitors from Boston, Massachusetts. Here, the pilgrim may pause to rest or to ponder American identity and Boston beginnings. Consider this final curiosity: Just eight miles away is a place, a hamlet actually, called Bunker's Hill.

Braintree

The intersection of the old Roman road from Colchester to St. Albans, one of the busiest in the land, and from London to Bury St. Edmunds made a natural choice for the establishment of the town now known as Braintree. In the earlier "Branchetreu"— meaning simply "town near a river"—the tribal chieftain Cunobelin, Shakespeare's Cymbeline, reigned up to the time of the Roman invasion. When Roman rule ceased after a period of over three hundred years, the area was occupied by the East Saxons, who gave their name to the county of Essex. With a favorable position in northern Essex between the rivers Brain and Pant, some forty miles from London, the steady growth of Braintree was a certainty.

The number of ancient inns extant evoke those ancient times when pilgrims stopped in Braintree en route to shrines at Bury St. Edmunds or Walsingham. The "Angel" sign, in Notley Road, imposes modern humor in its depiction of a winged angel with foaming pint, but it is the pint of beer which has earned the halo. Pilgrim traffic contributed to the growth of Braintree. So did wool.

Pub Sign in Braintree

A cloth-weaving industry, particularly of fine wool, was established in nearby Bocking early in the fourteenth century. One clothier's house in Bradford Street, now called Tudor House, has been restored and opened as a museum and relic of the active woolen cloth manufacture of medieval times. By 1389, silk weaving was established as well, and Braintree has been noted as a cloth-producing center ever since. Through changes in techniques, through waves of depression and unemployment—through centuries—Braintree has remained a textile town. Today, it is dominated by the huge Courtauld works which began here in 1816, having developed from a small family silk business into the world-known synthetic fabric manufacturing concern.

Tudor House Museum

Ancient government was served by a body of twenty-four citizens known as the "Four and Twenty." Although the origins of this little undemocratic oligarchy are unknown, it lasted over a century, from about 1565 to its final disappearance in 1716. According to just one of many theories which abound, the group is immortalized in the old nursery rhyme which alludes to "four and twenty blackbirds baked in a pie." It is a far cry from that

27

government to the present administrative setup which has merged the contiguous parish of Bocking with Braintree into a single unit. This must be a sensible and practical idea for, based on thirteenth-century accounts, pilgrims were just as uncertain as contemporary travelers where one ended and the other began.

While some ancient houses characterize the Bocking area, Braintree is bustling and modern. The Town Hall, in the Market Square, was built in 1928 on the site of a field on which William Pygot, a Protestant martyr, was burned at the stake in 1555. The Council Chamber of the Town Hall is decorated with murals depicting major events in Braintree history, and one portrays the victim being asked to recant as brushwood around him is about to be lit.

Another mural depicts the sailing of the Braintree contingent to the New World. The Reverend Thomas Hooker, originally from Chelmsford, is seen bidding farewell to the departing Dr. William Goodwyn of Bocking. Behind him stands John Bridge of Braintree, who is honored in the American Cambridge by a statue on the Common. Others kneel in prayer for a safe voyage.

In the entry hall is a model of the Lyon, the ship which carried the Braintree Company—actually 350 passengers recruited from various parts of Essex—to the New World in 1632.

The Braintree Company came to an earlier settlement of 1625 made up of indentured servants and called Mt. Wollaston, site of scandalous activities and a dubious past. Thomas Morton, the man responsible for setting up a Maypole at a place dubbed Merrymount, was expelled for his notorious festivities, and the unruly settlers were replaced by the hard-working, serious Braintree immigrants.

The Reverend Hooker arrived the following year to join his flock, having barely escaped summons and arrest by Archbishop Laud for his ardent preaching on Puritanism. But restless Thomas Hooker soon moved to Newetowne (now Cambridge) with many of his followers and, still dissatisfied, left there for the wilderness of Hartford, Connecticut. The remaining settlers first established a church and then established the town, which was incorporated in 1640 and called Braintree.

28

The Massachusetts city has much to be proud of in its heritage. Among the famous men to come from Braintree are two presidents of the United States, John Adams and his son John Quincy Adams, and John Hancock, President of the Continental Congress.

It is hard to imagine a place with closer, warmer feelings toward its mother town. Braintree, Massachusetts, presented a plaque to the Town Hall of Braintree, Essex, to commemorate the 750th anniversary of the granting by King John of the Braintree Market Charter in 1199. The weekly market insuring the town's position as center of the district still flourishes on Wednesdays, naturally enough, in Market Square.

The year 1199 also marks the founding of the Church of St. Michael the Archangel. The tower and east chancel wall of the church include Roman bricks, while most of the nave and tower is work of 1240. First enlarged in the thirteenth century, an ongoing series of renovations has resulted in a handsome building with harmonious adaptations. It has literary interest too in the fact that Nicholas Udall, vicar of Braintree from 1537 to 1544 and author of *Ralph Roister Doister*, is believed to have written plays which were performed here.

Braintree is surrounded by attractive villages. Coggeshall is particularly beautiful and historically important. Located on the old Roman route from Colchester to St. Albans, now the A20, it was a major cloth-making center in the fifteenth and sixteenth centuries and a lace-making center in the nineteenth. It has the remains of a twelfth-century Cistercian Abbey, the splendid Perpendicular church of St. Peter-ad-Vincula, and the beautiful, timbered, sixteenth-century house of the wealthy clothier, Thomas Paycocke, which is now preserved by the National Trust.

Great Maplestead has a Norman church, and Little Maplestead has a rare circular church (founded in 1340 by the Knights Hospitalers of St. John of Jerusalem), the smallest of five ancient round churches still existing in England. Finchingfield is one of England's most photographed villages, with seventeenth-century cottages set by the pond and village green, and a church of Norman origins in the background.

But it is easy to reverse completely the tourist activity of taking photographs. A large album, located in the Braintree Town Hall, was presented on the occasion of the celebration by Braintree, Massachusetts, of its tercentenary in 1940. Among such pictures of interest as the Rotary Club, Boy Scouts, and Women's Republican Club, is a copy of the deed of purchase of "Braintrey" from the Indians for twenty-one pounds and ten shillings. Thus, instead of taking photographs of Braintree in old England, the visitor can *look* at photos of Braintree in New England.

Brighton

The influx of August visitors to Brighton in England is a reversal of the exodus of August residents *from* Brighton in Massachusetts. The movement in both cases may be attributed in large part to throngs of holiday-bound trippers seeking the pleasures of sunny shore resorts.

The tradition which nurtured the stylish seaside resort of Brighton in Sussex and made it a standard for measuring seaside resorts elsewhere can perhaps be said to have emanated from Dr. Richard Russell, the "inventor" of sea bathing. Dr. Russell had been prescribing to his patients the sea-water cure with such enormous success that he moved his practice to Brighton in 1754 to further foster his belief in sea bathing as a health-inducing activity. His prescriptions transformed the poor fishing village of Brighthelmstone into a fashionable watering place. Salubrious sea water, it was proclaimed, could cure a range of diseases from asthma and rheumatism to consumption and cancer. But surely, it must have been its promise to renew sexual prowess and vitality that had no small part in the guaranteed success of this fashionable form of medical therapy.

After Dr. Russell published a thesis, in 1750, on the meritorious effects of sea water—both for bathing in *and* for drinking—Brighthelmstone sea water was bottled and sold in London to those who could not get to Brighton, much as today tins of Cape Cod air are sold, presumably for those who cannot leave their sweltering cities. Dr. Russell's epitaph bears the quotation from Euripides which must have been his motto in life: "The sea washes away all the ills of mankind."

In the early history of the new pastime of sea bathing, a device known as the bathing machine was used. A small shed on wheels served as a dressing room, or more accurately as an undressing room, for men generally wore no bathing costumes at all until about 1865, while women might be clothed in a kind of nightgown. The participant remained in the shed to be immersed

31

Royal Pavilion

as the chariot was drawn out into the water by horses. And perhaps a viewer or two remained on shore with a telescope to peruse splashing, naked figures. Brighton never adopted the modest precaution of a canvas hood over the steps of the bathing machine to conceal the bather—a practice that was characteristic of less scandalous shore resorts such as Margate, Weymouth, or Scarborough. By about 1750, attendants known as "dippers" and "bathers" established their personalities on the scene with their function to make certain that their charges were properly immersed.

The fashionable Sussex seaside resort attracted many fashionable visitors over the years. The historian Edward Gibbon came in 1781, just after publication of his third volume of *Decline and Fall of the Roman Empire*. Charles Dickens stayed on several occasions and wrote *Bleak House* and *Dombey and Son* here. Thackeray visited and included a description of "brisk, gay, and gaudy" Brighton in his *Vanity Fair*. Dr. Samuel Johnson was another distinguished visitor. Among the artists are such notables as Sir Joshua Reynolds, John Constable, and J.M.W. Turner.

But it was the arrival of the Prince of Wales in 1783 that was to brighten the character of Brighton and change it irrevocably. From his very first visit, the Prince (later King George IV) was enchanted with the town and returned regularly. The villa that he required was continually enlarged or altered, over a thirty-five year period, to accommodate and reflect his brilliant social life. The ultimate design for Regency revels is credited to the architect John Nash. It was his plan, when the Prince a stately pleasure-dome decreed, that brought about the realization of an oriental fantasy. The English imagination, captivated by Eastern splendors, now had its own splendid rendition in the minarets and pinnacles and onion-shaped domes of the fantastic Royal Pavilion.

Queen Victoria, however, did not find it suitable to her need for privacy, and the Royal Pavilion was sold to the town in 1850. The elegant eccentricities of the entire estate have been a delightful center for the public ever since.

The glass-domed Royal Stables were converted into a concert hall. The riding school, known as the Corn Exchange, is used for flower shows and other exhibitions. The former indoor tennis court is the present Art Gallery. But the Royal Pavilion, with its sensational Indian effect, is at the center of the estate and indeed of all that is associated with the panache of Brighton.

The interior of the Royal Pavilion is no less sensational. Among the lavish interior eccentricities which may be viewed and enjoyed are the banqueting room (with exotic plantain tree ceiling and rich dragon decorations); the kitchen (full of copper pans, roasting spits, and stylized palm tree columns); the music room (with stylized lotus-like chandeliers arranged around the circular room); and the king's apartments (with hidden door leading to a bedroom above).

But Brighton was not always a gay and sophisticated pleasure resort. The Brighton Museum exhibits a past much older than Regency. Skeletal remains and artifacts from recent archeological excavations give evidence of the earliest inhabitants of Brighton some five thousand years ago. And a nearby prehistoric hill fort called Hollingbury Camp further attests to a much earlier civilization.

Just as its use as a resort is a relatively modern phenomenon, so its name is a modern adaptation. It seems probable that the name of Brighton, which derives from Brighthelmstone, may be further traced to a St. Brightelm. In one variation of his story, he is an Anglo-Saxon bishop who accompanied the Saxon army and died in battle in 693. The popular and current form of the name came into use around the end of the eighteenth century and into official use in 1810. It seems highly likely that Brighton, Massachusetts, incorporated in 1807, and now part of Boston, was named for that archetype bathing resort because ancestors of several early American settlers came from that area.

Brighton today is a modern residential town, a dormitory for commuters to London, a center of light industry, and an academic mecca which can boast of its University of Sussex, as well as a seaside resort. A long way from neolithic man are the monolithic traffic jams created by travelers to Xanadu. In a curious reversal, from the legalized speed limit set in 1896 of fourteen miles an hour (an occasion still commemorated each year by a procession from London to Brighton of pre-1905 cars), motorists today can often average four miles an hour on fine Sunday summer evenings. The traffic congestion problem is inevitable because of its location—London's nearest point on the English Channel, a distance of only fifty miles.

But also because of its location on the Channel, Brighton has never really been free from fear of invasion. Most recently, it has been invaded by modern developments of flats, shops, restaurants, and office buildings. One area, however, known as the Lanes, preserves the character of the old town and makes shopping in Brighton a particular and pleasurable attraction. The Lanes are a maze of alleyways crowded with shops purveying antiques and rare items and retaining the original flavor of seventeenth-century Brighthelmstone fishermen's cottages.

The seafront sets the holiday tone with its two piers (Palace Pier offers entertainments, fun rides, and refreshments), broad promenade, and south-facing beach.

Although Brighton was badly bombed during the war, the Royal Pavilion was not touched because Hitler intended to use it for his personal headquarters. It remains the cynosure and symbol of romantic potentialities. Perhaps the fiction of Jane Austen is indeed reality, as when a character in *Pride and Prejudice* suggests that "a visit to Brighton comprised every possibility of earthly happiness."

Bristol

It is small wonder that there are fifteen towns called Bristol in the United States. For hundreds of years before American settlements were founded, Bristol was one of the great maritime centers of the world and the largest and busiest port in England after London.

A year before Columbus landed on the American mainland, John Cabot set sail in 1497 from Bristol, England, to discover Newfoundland and North America. A statue has been erected on the quayside in the harbor area to honor the great explorer. The new continent developed as many English emigrants sailed from the active port of Bristol to colonize settlements on the Atlantic seaboard: Virginia, Maryland, Pennsylvania, Connecticut, New Hampshire, Massachusetts, Maine. . . .

Located on the River Avon, 113 miles west of London, Bricgstow—the settlement by the bridge—became established as a port in Saxon times and flourished as a port from the tenth century onwards. The city grew up around its harbor and became an important enough commercial center to be made a county in its own right in 1373 by royal charter of Edward III.

During the seventeenth century, ships sailed from Bristol laden with English people of all classes who emigrated to the new land for all reasons. Among those who left because of religious dissent was a leading Bristol Quaker, William Penn, founder of Pennsylvania. An enormous increase in trade included importation of tobacco from Virginia and sugar from the West Indies through Bristol, and Bristol ships carried slaves to plantations.

Bristol is today an exciting city with a great many sights to entertain, educate, and occupy tourists. The Theatre Royal, built in 1766, is the oldest existing theatre in England and the home of the Bristol Old Vic Company. The more recent Hippodrome (1900) has a seating capacity of two thousand and the largest provincial stage in Britain.

The University of Bristol, founded in 1876, was the first in the country to include drama as a subject in a degree curriculum when it opened a special department in 1949. The impressive University Tower, opened in 1925 by King George V, houses a ten-ton E-flat bell called Great George, with a sound that heralds the majestic interior.

Near the university is the Bristol Museum and Art Gallery with a diverse collection that gives prime space to a replica of the plane known as the Bristol Boxkite, built for the film *Those Magnificent Men in their Flying Machines*. But much remains in the city itself from days of old, beginning with the cathedral.

Originating as an Augustinian monastery in 1148, the cathedral was already three hundred years old when Cabot sailed for America and made Bristol known throughout the world. It retains the original Norman chapter house, distinctive fourteenth-century vaulting in the choir and aisles, soaring arcades, fifteenth-century cloisters, and misericord seats in the choir carved with such subjects as dancing bears, pig-killing, and wrestling. A Saxon sculptured coffin lid in the south transept is one of the oldest relics in the cathedral.

Another sight to please antiquarians is the medieval entrance arch which led to the courtyard of the thirteenth-century St. Bartholomew's Hospital. The arch is at the bottom of Christmas Steps, a colorful passage with intriguing Georgian shops and a Dickensian look, designed originally to link two main streets.

At the top of Christmas Steps are the Chapel of the Three Kings and the adjoining Foster's Almshouses, both initially funded by a salt merchant in the fifteenth century. The present almshouses date from 1861, but Colston's Almshouses, set back from St. Michael's Hill, were founded in 1695.

St. Mary Redcliffe, praised by Elizabeth I as the "fairest, goodliest and most famous parish church in England," contains a painted wooden statue of the Queen. Beneath its tower is the Chapel of St. John the Baptist, known also as the American Chapel because American Friends of St. Mary Redcliffe restored and furnished it.

In nearby College Green, much of the city's ancient history is displayed in the decor and paintings of the attractive and new

Council House, completed after World War II. Across College Green is the Lord Mayor's Chapel, with an exquisite interior, a remnant of the medieval Hospital of the Gaunts founded in 1220. And for a change of subject, behind the chapel and incorporating the thirteenth-century cellars of Gaunts' Hospital, is Harveys Wine Museum—a reminder of the days when wine was a major import and the basis for later successful firms. The wine museum is run by a firm established in 1796, famous today particularly for Harveys Bristol Cream sherry.

For an idea of what homes of wealthy merchants were like, the Georgian House is a museum which opens to view the three floors of the former home of John Pretor Pinney.

Four streets—High Street, Corn Street, Wine Street, Broad Street—make up the core of the old Saxon walled town and are good grounds for exploration, occasionally revealing an American association. The Church of St. Nicholas houses an altarpiece triptych by Hogarth, but in the crypt is the tomb of John Whitson, who promoted Pring's voyage to New England. Martin Pring, buried in the nearby fifteenth-century St. Stephen's Church, sailed to New England in 1603 in search of sassafras and other trade opportunities and anchored in Whitson Harbor, later called Plymouth by the Pilgrim Fathers.

The Nails, Corn Street

In St. Stephen's Church is a monument to the discoverer of Plymouth Harbour, U.S.A.

Not to be missed in this old quarter is the eighteenth-century Corn Exchange where merchants completed their transactions. Cash would be placed outside on four flat-topped, bronze pillars known as the nails, thus giving rise to the phrase, "paying on the nail."

The city is a seemingly unending source of surprising sights. Near the main shopping center known as Broadmead is the New Room, the first Methodist chapel in the world. Opened in 1739, visitors may still see the original galleries, the dark mahogany box pews which contrast with the white-paneled walls, and the double decker pulpit from which John Wesley preached his rousing sermons.

Bricgstow began in what is now Castle Park, dominated by the shell of St. Peter's Church. Severely bombed during the war, the area has been excavated to reveal Saxon houses. The River Avon flows alongside Castle Park and under Bristol Bridge.

A further American association is disclosed in the fashionable suburb of Clifton, just over the Clifton Suspension Bridge. The former spa town, its spring of warm water bubbling out from the Avon mud below St. Vincent's Rock, with a profusion of Georgian and Regency architecture, once vied with Bath as an elegant health resort. Clifton College flies the American flag on Independence Day to commemorate wartime service as General Omar Bradley's headquarters.

California

Totally unlike the large state on America's west coast, the English California is a tiny, unknown, and sparsely-populated place on the East Anglian coast characterized, not by blue skies, but by windy, arctic weather. Nor is it a mecca for millions. Indeed, it is reminiscent of the grand and glorious state of California in a single attribute: its name.

Even the British are generally unaware of the existence of California in their own country. On England's dramatic east coast of Norfolk, facing the winds of the North Sea, is a stretch of sandy beach and crumbling cliffs where erosion has caused whole towns to topple into the water. The coastal belt north of Great Yarmouth is nevertheless exquisite for its unspoiled and rugged landscape punctuated by gale-twisted trees. The area is a bird-watcher's paradise and haven for nature lovers.

One of a series of small settlements north of Yarmouth and Caister, California diverged from historical procedure when it chose its intriguing name. It reversed the traditional practice of bequeathing names of old British cities to New World settlements and selected the famous American place where gold was found in 1849 as a name source for Britain.

But California must be twice as popular in Britain where, to compound the improbable, two Californias actually exist. In addition to the small village in England, another is located in Scotland about halfway between Glasgow and Edinburgh.

The Scottish village made its first official appearance in the Valuation Rolls in 1882 when it was a mining community. Iron pits and coal mining, together with spoil tips, gave the locals the idea that the landscape looked like the California gold-rush scene, and they applied a ready-made name to the village.

But primitive miners' cottages have been superseded by council houses with electric lights, cold running water, and bathrooms. Few residents still work as miners, and most travel to the larger towns to earn a living. With a population now of a mere two to three hundred, the church has recently been demolished. But the community manages to survive with a congregation that meets in the Village Hall, a primary school, and various small village activities.

Indeed, gold is also the reason for the existence of California in Norfolk. When the little place on the East Anglian coast was settled in 1849, it too was named for a gold rush, albeit a much smaller variation of the well-known one on the west coast of another continent. Erosion of the cliffs near Caister released some sixteenth-century gold coins at the base of the Scratby cliffs, just at the site where migrating fishermen from Winterton had set up a colony. The year was 1849. The parallel was drawn. Would not the gold-rush state of California supply a suitable name for the new colony? The beachmen were themselves, after all, in search of a golden opportunity.

The migrants had come to this site to be nearer the source of their livelihood, the fishery at Great Yarmouth. But they had in mind another and more lucrative motive—to salvage vessels from the sandbanks around Yarmouth where, in the eighteenth and nineteenth centuries, sailing vessels were regularly stranded by bad weather. Salvage groups, known as beach companies, established themselves all along this treacherous part of the East Anglian coast to extricate floundered ships for a fee.

In addition to the sense of humor evinced in the name they chose, the company showed a sense of practicality by choosing a cliff lookout. With a better view of the sea, they gained a distinct advantage over rival beach companies.

They further outstripped their competitors in the business of salvage work and lifesaving when they bought a lifeboat which enabled them to work in rougher seas. Using an inventive public relations tactic, they applied for and received permission to use the Prince Consort's name; the added prestige of the *Prince Albert* lifeboat insured success. With each salvage procedure bringing in an average of about £100 per vessel—more valuable cargoes paid more—the gold continued to stream in until changing times introduced modernized ships.

It was a brief heyday. Some fifty years after its founding, California nearly ceased to exist when, in addition to safer

41

shipping, erosion began a destructive course. A gale of 1897 washed away ten yards of cliff. The lifeboat had been sold in 1894, and residents had begun migrating around the turn of the century to nearby Yarmouth. During World War I, beach company buildings, frontage cottages, and the California Tavern, were demolished for fear that the the crumbling coastline would tumble them into the sea.

But the colony was resurrected with another gold rush that came in the 1960s, this time in the form of golden sands. Tourists began to flock to California for seaside holidays, renewing its prosperity. Symbolically, the California Tavern was rebuilt and given a second life.

Nowadays, California awakens from a dormant winter sleep to become a summer spectacle of tents, trailers, holiday bungalows, small amusement centers, miniature golf, ice cream trucks, and fish and chips. It revives as a swinging community which boasts of a week-long festival of country and western music attracting large numbers to this resort some 127 miles northeast of London.

Instead of beachmen overlooking the sea from their clifftop surveillance site, a tenting community overlooks a wide beach dotted with warning signs, "Caution Beware Cliff Falls"—a reminder of the reason for the former settlement's demise. And even on a rare hot summer day, sun bathers plant their colorful plastic screens on the beach as protection against the cold North Sea wind—a reminder of the gales that caused vessels to flounder and brought about the initial founding of California.

The association with gold suggests also the varied treasures to be discovered in the surrounding area—in the historic town of Great Yarmouth, the inland area of low-lying land known as the Norfolk Broads, at Caister-on-Sea with its Roman excavations and fifteenth-century castle of Sir John Fastolf, and the county city of Norwich with its splendid cathedral and its many museums and art galleries.

Many would deem it appropriate for the annual California music festival to feature as its theme a variation on the well-known song: California Here I Come—to England.

Cambridge

Picture this sunny scene: a river setting with punts or sculls plying their way upstream, students reading or relaxing on the grassy embankment—perhaps watching a boat race, perhaps watching joggers—university buildings in the background, attractive footbridges over the river, visitors or shoppers taking a restful break in the friendly atmosphere of what is unquestionably a university town. Cambridge, yes! But which one?

When the American settlement of Newetowne was selected in 1638 by the ministers and leaders of the community, most of whom were educated at Cambridge University, England, to be the new seat of learning, its name was changed— reasonably—to Cambridge. It was expected that the newborn infant would grow and emulate its parent in every way. The college itself was named for a man who bequeathed his entire library at his death in 1638. An American in England wishing to pay homage to John Harvard might visit the Harvard House in Stratford, the early home of his mother, or the Harvard Chapel in Southwark Cathedral, London, where he was baptized. Founded in 1636, the new institution was well on its way toward fulfilling its potential, toward attaining the status of the older counterpart, when John Wilson wrote in 1704:

And as old Cambridge well deserved the name,
May the new Cambridge win as pure a fame.

It is not possible to ascertain the origins or precise date of the founding of old Cambridge University. In the twelfth century, the word *universitas* referred merely to an organized body of men, a corporation. Young men simply came in the Middle Ages to the ancient religious houses of greater abbeys to be educated and lodged with a master or in private quarters. What is known is that in 1209 a migration of scholars from Oxford to Cambridge took place. They selected Cambridge perhaps by chance,

43

perhaps by reputation of existing schools and teachers; but a university has been in existence at Cambridge from that date.

Peterhouse, the earliest Cambridge college, was founded in 1284. St. Peter's Church and two neighboring hostels were turned over to Bishop Hugh de Balsham for his body of secular scholars to live together as students of the university. The college was ravaged by fire in the fifteenth century, and hardly any of the original buildings remain. When the poet Thomas Gray lived at Peterhouse in 1756, he fixed an iron bar to his window on the north side of the main court. In case of fire, a rope ladder could be suspended from the bar, which is still visible. His phobia made him the unhappy butt of an undergraduate prank. After a false alarm with cries of fire caused him to slide down the rope ladder in his nightshirt and into a tub of cold water, Gray transferred to Pembroke, across the road.

King's College was founded by Henry VI in 1441. Its many fine features are dominated by one of the most magnificent examples of Perpendicular architecture in England—King's College Chapel. A very large area of brilliantly colored glass imparts a feeling of lightness to the vaulted roof. Thus the poet John Betjeman describes the network of tracery and fan vaulting: "With what rich precision the stonework soars and springs to fountain out a spreading vault—a shower that never falls." When the voices of the world-famous choir of King's College Chapel fill the void, the result is inspirational. Wordsworth's sonnet speaks of

> that branching roof
> Self-poised, and scooped into ten thousand cells,
> Where light and shade repose, where music dwells
> Lingering—and wandering on as loth to die;
> Like thoughts whose very sweetness yieldeth proof
> That they were born for immortality.

Adding to the riches of the chapel is *The Adoration of the Magi* of Rubens which forms the altar-piece. No, it is not possible to assimilate the splendors in one brief visit. A member of the chapel staff recalls an American couple who explained that they had flown to England just to visit King's College Chapel, Cambridge. They gazed at the lofty vault and ecstatically left after a two-day jaunt across the Atlantic, having paid the ultimate tribute.

King's College Chapel

Queens' was founded by two queens, Margaret of Anjou and Elizabeth Woodville. Queen Margaret founded it in 1448 while her husband, Henry VI, was busy with the founding of King's. Her purpose was to honor women. The need to educate them was not appreciated until 1869 with the establishment of Girton, a college for women. Edward IV's queen, Elizabeth Woodville, took over the sponsorship in 1465 after the defeat of Henry VI in the Wars of the Roses. Her portrait hangs in the hall behind the president's chair. The Cloister Court which she built is one of the most beautiful in Cambridge; a spectacular Tudor building on the north side of the court is the half-timbered President's Lodge. In the southwest corner of the principal court is Erasmus's Tower, aptly named since that is where the great humanist stayed when he was at Queens'. The wooden Mathematical Bridge, built in 1749 without a single nail, gives a fine view of the college.

Trinity is the largest. Henry VIII founded it in 1546, a year before his death. His statue is on one side of the Great Gate. In the hall, notable for a fine hammerbeam roof, is a famous portrait of Henry VIII by Holbein. The Wren Library with carvings by Grinling Gibbons is a treasured attraction both for its architectural design and for its contents. To Trinity belong such names as Dryden, Byron, Macaulay, Tennyson, Housman, Sir Isaac Newton, and John Winthrop of New England fame.

St. John's, the second largest in Cambridge, has a magnificent turreted gatehouse, a splendid Tudor hall with hammerbeam roof, and an enclosed bridge with traceried windows over the River Cam, the famous Bridge of Sighs, named for its counterpart in Venice.

Magdalen contains the library bequeathed to it by Samuel Pepys. In the Pepysian Library, among some three thousand books of value and interest, was discovered the famous *Diary* that was later transcribed from the shorthand of the author's private system and published in 1825.

In the famous garden of Christ's College is the mulberry tree which makes it famous, for it is associated with a student of renown who was at Christ's from 1625 to 1632. John Milton is said to have written *Lycidas* under this legendary tree.

Founded in 1497 on the site of a Benedictine nunnery, Jesus College has many old and beautiful bits of architecture, especially the thirteenth-century St. Radegund's Church, now the chapel of the college and updated with stained-glass windows by Burne-Jones. In the library of Jesus is the first

Bridge of Sighs, St. John's College

edition of the first Bible ever printed in America—in fact, in Cambridge. A Jesus man, John Eliot, whose mission it was to convert the Indians, had it printed in 1663 in the language of the Mohicans.

Emmanuel, founded in 1584, was one of the later colleges. It was a strongly Puritan college and the source of many emigrants to America, including John Harvard. A window to the memory of John Harvard was placed in the Wren Chapel in 1884, at the tercentenary celebration.

It is not possible to uncover all that Cambridge has, not even if one lives there. Beyond college gateways are peaceful courts, beautiful gardens, and quiet cloisters, all enhancing those fine examples of college architecture. The area between the great colleges and the River Cam, known as "the Backs," is a great delight, particularly to picture takers or to paddlers who may hire boats. The forty acres of Botanical Gardens had as its original purpose to show uses of plants in medicine. Of the many museums, the Fitzwilliam is one of the finest in the country; its collections of paintings, engravings, books, and other valuables should keep a visitor enthralled for many, many days at least.

Colleges and college life dominate, but the city is much older than its university. Because of its site, Cambridge had to be important. Located on the River Cam, it provided the only good possibility for crossing by means of a ford and later a bridge. Around the firm ground which became Cambridge, stretched the undrained fenlands—swamps, streams, and lakes—which separated East Anglia from the rest of the country. Thus, Cambridge was an oasis even some two thousand years ago.

When a Roman fort was established in the year 43, it replaced an earlier settlement. Clearly, the Romans appreciated the need to guard the fort from a position on the high ground to the west of the river which they knew as the Granta. By the second century, the Roman town had a grid pattern of streets, and Roman roads passed through or near Cambridge.

In 875, Danish invaders occupied the town called Granta Bridge (by then the ford was bridged) and made it their local army headquarters in 921. The ever-changing name evolved from Grantebrige, Cantebrigg, Cauntebrigg, and Caumbridge to its modern form, which came into use in the sixteenth century.

In the tenth and eleventh centuries, Cambridge was very important because of its location at the head of a river, with access to the sea via King's Lynn, and as a point of communication between East Anglia and the rest of England. The Church of St. Benet was built in this period, and the original tower of 1025 is still standing—the oldest building in town and one of the most complete Anglo-Saxon buildings in the country. The church is connected to Corpus Christi College by a sixteenth-century brick passageway.

All that remains of the castle built by William in 1068 after the Norman Conquest is the mound on Castle Hill.

The early twelfth-century round church, the Church of the Holy Sepulchre, is the oldest of five round churches extant in England. Its design was based on the plan of the Holy Sepulchre in Jerusalem and influenced by the returning crusaders.

Great St. Mary's in King's Parade is the university church, and its tower is well worth climbing for a great view of Cambridge.

Little St. Mary's, near Peterhouse, is a little gem which contains the interesting monument to Godfrey Washington. The stars-and-stripes coat of arms of this clergyman recalls the American flag given by a well-known member of his family.

Near this Place lyeth the Body
the Late Rev.d M.r GODFREY
WASHINGTON of the County
of York, Minifter of this Church
and Fellow of St.Peter's Colledge
Born July the 26.th 1670
and Dyed the 28.th day of Sept.r
1729.

The Washington Memorial in Little St. Mary's

Not limiting itself to history, Cambridge even has an amusing bit of lore attached to it which involves Thomas Hobson. This wealthy town figure, many times mayor, kept forty horses in his livery stable for hire by a system of rotation. A customer could hire any horse, provided it was the one Hobson chose. He died in 1630 but achieved immortality in the phrase, "Hobson's choice."

Cantabrigians on both sides would surely agree that Cambridge is inexhaustible and that, taking both sides into account, Cambridge must be one of the best known and prestigious names in the world.

Chatham

Chatham, contiguous with the Dickensian city of Rochester (few people know where Rochester ends and Chatham begins), also (like Rochester) has Dickens associations. It was in Chatham, thirty-two miles southeast of London, that Charles Dickens first came to know his beloved county of Kent. He was brought here as a child when his father worked as a clerk in the dockyard that was founded in 1547 and employed some fifty shipwrights by the end of the sixteenth century. But the dockyard that built such well-known ships as Nelson's *Victory* was shut down by the Royal Navy in 1984 and has since been transformed into a living museum. Rope making is displayed in the Ropery, and sail-makers are at work in the very loft where Nelson's flags were made. Naval guns and historic buildings are among the exhibitions contained within an eighty-acre site.

On a hill overlooking the docks is a vast Georgian fortress known as Fort Amherst with a complex of tunnels, storerooms, and living quarters, as well as a museum on the ground floor. A barrack room on an upper floor recreates the living conditions of Wellington's soldiers, and scenes from the history of the fort are re-enacted during the summer months.

Upnor Castle, on the River Medway just beyond Rochester, was built in 1559 to protect Chatham Dockyard from attack. But the neglected and unprepared castle failed to stop the attack that eventually came in 1667. The Dutch fleet sailed right in and captured the British ships anchored on the river. Now open to visitors, the castle houses an exhibition of that Dutch raid. But the most recent raid came in 1941 when two bombs fell in the garden of Upnor House.

Upnor itself is a delightful little village of three streets, lined with weatherboarded cottages. The narrow High Street (in which two pubs beckon) slopes down to a pleasant stretch of the Medway with a marina and with access to the castle. Here the Medway makes a final bend before the estuary opens out to the sea. It all serves as a reminder of the strategic importance of Chatham and the river in centuries past.

Although the life of Dickens began in Portsmouth, where he was born in 1812, it continued in Chatham at 11 Ordnance Terrace (near the railway station) and later in cheaper lodgings in St. Mary's Place (demolished in 1943). He died at nearby Gadshill in 1870 in the area where he had spent his happy childhood years.

It was from Chatham that he often took long walks with his father, exploring the countryside and ending up by Gad's Hill, scene of Falstaff's revels. There father and son admired a particular rose-brick country house. In later life, rich and famous, Dickens was able to purchase the very house he had coveted as a child when his father was a pay clerk in the Chatham dockyard. He lived at Gad's Hill Place for the last twelve years of his life and died there on 7 February 1870.

Chatham, a name probably based on the Teutonic word describing a forest settlement, came into use after the Romans (who would have held back attacks from Teutonic tribes) departed in the fifth century. The Jutes invaded, initially for plunder; but they later settled in the land, recording one of their new communities as Cetham, Caetham, and eventually Chatham.

The town expanded as the Medway's importance to the navy grew in the mid-sixteenth century. Dockyards and barracks developed, and dockyard laborers and artisans came to live in the cheap housing available in Chatham.

In America, several colonies were named Chatham by settlers known to have originated from England. In Massachusetts, the former settlement or district of Manamoit was established as a town on June 11, 1712, with the new name of Chatham, probably derived from emigrations from Kent.

In New Jersey, the Seelys, who can be traced to New England in the 1630s and then back to England, settled in the rolling hills of the northern part of the state in 1772. Other settlers, descendants of Englishmen from Connecticut or Long Island, joined the community. At a time of political unrest and turmoil elsewhere in the colonies, including the Boston tea party, the inhabitants called a meeting on the 19th of November 1773 to choose a proper name for their settlement. They had outgrown the Indian designation of Minisink Crossing. John Day's Bridge and Caster's Bridge, names of the 1760s, preceded the prosaic name of "On Passaic River," which might correctly apply to any of a number of villages on the meandering river. The unanimous vote produced a name which was not an Indian word, or the name of an early settler, or a nostalgic tribute to England. Chatham honored an Englishman embroiled in political battles between the mother country and the American colonies who had opposed the tea tax and the Stamp Act of 1764 on grounds that the kingdom had no right to tax Americans who were not represented in Parliament. It honored William Pitt, Earl of Chatham, England.

Chelmsford

Recorded history traces Chelmsford back to the middle of the first century when it was the site of the Roman Caesaromagus on a fordable part of the river Chelmer.

Actually, two rivers, the Can and the Chelmer, divided the town into three sections. When the bishop of London built the first bridge over the river Can in about 1100 during the reign of Henry I, the town inevitably increased in prestige and scale and never really stopped growing. By 1189 Chelmsford was the assize town. By 1200 it had a market and a fair. So vital was the change, that in the coat of arms is a representation of that historic bridge of three arches. The present Stone Bridge dates from 1787.

Situated just twenty-nine miles from London, and just a half hour away by train, Chelmsford has developed into an industrial center with several important factories. In particular, it is the birthplace of the radio industry. A company formed by Marconi started operations here in 1897. After successfully spanning the Atlantic by wireless signal in 1901, expansion was rapid. Today there is concentration on all sorts of electronic equipment dealing with broadcasting and telecommunications.

In spite of industry, this cathedral town and county town of Essex manages to maintain antiquities and characteristics which make it special. The Chelmsford and Essex Museum, in the Victorian mansion of Oaklands Park in Moulsham Street, contains such varied displays as Bronze Age implements, Roman pottery and artifacts, medieval English arts, British birds, and a collection of paintings which includes a Turner and landscapes by Constable and Gainsborough.

The contrasting and rushing flow of contemporary life is in the town center with its busy stores and offices, new pedestrian shopping center, buildings of architectural interest, and bits of old history. Anthony Trollope worked on his novels in the Saracen's Head, located on the High Street near Tindal Square. Sir Nicholas Tindal is honored by his statue in the Square. Born in Chelmsford in 1776, this local boy made good by working his way up to finally become a Chief Justice before he died in 1846.

But there is a memorial of another kind to a local boy who made good in America. Thomas Hooker, a minister of St. Mary's in Chelmsford, who later left for Braintree, emigrated to New England at about the time of the Pilgrim Fathers. The American Chelmsford was named for its older counterpart, the home of Thomas Hooker and other early settlers.

The church in which the Reverend Thomas Hooker had been minister, now a cathedral, is almost hidden behind shops along Tindal Square and overshadowed by offices of County Hall. The typical Essex parish church attained cathedral status in 1914 when Essex became a separate ecclesiastical diocese. Built in the fifteenth century in the Perpendicular style and rectangular plan of the period, St. Mary's is notable for a unique fifteenth-century fan arch in the north wall of the chancel and for seventeenth-century Dutch altar rails. But its tower shows Norman traces, although it was rebuilt in 1424.

Tindal Square

Much that is new includes the Bishop's throne (1922), the life-size statue of the first Bishop of Chelmsford, John Edwin Watts-Ditchfield (1931), the high altar (1931), and medieval- style coloring of the sanctuary and chancel roof (1957). A modern St. Peter squats outside on the southeast corner, a modern fisherman holding a contemporary house key. The porch was beautifully restored in 1953 as a memorial to United States airmen stationed in Essex in World War II. Stained glass represents the seal of the United States of America, and of the United States Air Force, as well as the Washington family arms; the great-great- grandfather of George Washington was an Essex clergyman. An inscription in the glass explains that the porch was "enriched and beautified by Essex friends of the American people" and indicates a continuous friendship.

St. Peter on Southeast Corner of Chelmsford Cathedral

Chelmsford is within easy reach of London and within easier reach of beautiful little villages. The unspoiled village of Little Writtle, two miles west of Chelmsford, with its old houses set around a triangular green with a pond at the apex, is perhaps the most tranquil. Surprisingly, it was more important than

Chelmsford at the time of Domesday; but when the bridge was built in Chelmsford, the highway was diverted and so was Writtle's prosperity.

Ingatestone, six miles southwest on the Chelmsford-London Road, is a pleasant village with a sixteenth-century inn and a fine mansion known as Ingatestone Hall, erected in 1565 by Sir William Petre at the time of the Dissolution of the Monasteries. One of its features is a priest's hiding hole concealed in an upper story. Elizabeth I was lavishly entertained at the Hall, and William Byrd, the Elizabethan composer, was a frequent visitor. More recently, the house was the setting for Elizabeth Braddon's novel, *Lady Audley's Secret.*

Yet Chelmsford, because of easy access, inevitably gets the spillover from London. It is a commuter's community like its American counterpart, which is also within easy access of the big city. If that similarity were not striking enough to call up images of New England, road signs point to such familiar names as Springfield, Waltham, Billericay, Maldon, and Braintree.

Dartmouth

The old and picturesque town of Dartmouth, situated on a hillside just inside the mouth of the lovely River Dart, has a magnificent, sheltered, deep-water harbor and a dramatic setting on the South Devon coast of the English Channel.

The town began life in the twelfth century. From here, the Second Crusade departed in 1147; and the Third Crusade, led by Richard the Lion Heart, in 1190. But the most recent and largest gathering for a military operation occurred in 1944 when 485 ships of the United States Navy used this important port of embarkation for the Normandy invasion.

Long before this modern crusade, in medieval times, the town grew rich on piracy, privateering, and a thriving wine trade with Bordeaux. Chaucer visited Dartmouth in 1373 as a customs officer and is believed to have met John Hawley, an importer of French wine and the greatest of all merchants and shipmasters of that thriving medieval port. He may be the wily character that Chaucer alluded to in *The Canterbury Tales*:

> A Shipman was ther, wonyng fer by weste
> For aught I woot he was of Dertemouthe.

One of the best memorial brasses in the country is in the chancel of the Church of St. Saviour, near the quay. An armored John Hawley, who died in 1408, is depicted with his two wives, Joan and Alice, on either side. The church is also notable for its fifteenth-century screen with intricately carved friezes and the elaborately carved fifteenth-century stone pulpit on a slender pillar.

Elizabethan and Jacobean houses can still be seen on the quay and elsewhere to testify to the town's wealthy past. Particularly appealing is the Butterwalk. The seventeenth-century colonnaded row houses consist of shops below and timbered living quarters above, which, supported by granite columns, project out over the sidewalk. Number 12, now a chemist's shop, has a fine plaster ceiling of the Tree of Jesse. The Borough Museum is at Number 6. Built in the reign of Charles I in the 1630s, the Butterwalk was restored after enemy bombing during the war caused some destruction.

The Butterwalk

Dartmouth's location gave it a significant role in the exploration and settlement of the New World. Among the explorers based here at some point in their careers were Sir Walter Raleigh, Humphrey and Adrian Gilbert, and John Davis.

On the cobbled quay known as Bayard's Cove is a monument to what is perhaps the best-known voyage of exploration. The Pilgrim Fathers paid an unexpected visit to Dartmouth in August 1620 for eight days of repairs to the *Speedwell*. Apparently the work done was inadequate because the *Speedwell* and its accompanying ship, the *Mayflower*, had to put in again at Plymouth for further repairs. (The *Speedwell* was eventually abandoned.) Thus, because it was not the last port of call for the colonists, the town was deprived of the opportunity of being immortalized in the New World with a "Dartmouth Rock."

From the quayside, an excursion can be taken from the estuary of the River Dart to Totnes, a twelve-mile journey. The journey goes past sleepy villages with such obscure names as Dittisham and Stoke Gabriel. Points of interest along the route include the Royal Naval College (which has been training naval officers since 1905), the large shipyard at Noss Point, the Greenway House home of the late Agatha Christie, and Sandridge House built by John Nash in 1805.

Dartmouth Castle, built in the 1480s in the reign of Henry VII, faces a similar castle on the opposite shore across the estuary at Kingswear. The twin castles were designed to guard the narrow entrance by a thick iron chain that could be stretched across the river to prevent enemy ships from entering. Kingswear Castle is now a ruin and not open to the public, but Dartmouth Castle remains very much as it was when Henry VIII, fearing an attack from the Continent, further strengthened the town's defenses in 1537.

Next to the castle stands the Church of St. Petrox, a simple seventeenth-century building, bare and without chancel or transepts but with a Norman font and a plethora of plaques and brasses. Together with Dartmouth Castle, it forms an impressive scene on the west side of the Dart.

In the tropical Royal Avenue Gardens near the quay is a monument to Thomas Newcomen of Dartmouth who invented the first practical steam engine, used locally in West Country tin mines. The museum next to the gardens, as if to atone for the slipshod technology committed on the *Speedwell*, is dedicated to the founder of the Industrial Revolution. On view is one of his original atmospheric pumping engines of 1725, the oldest in the world.

This delightful Devon town invites the meanderer to wander and take in the history and architecture, the pubs and pleasures, of a most satisfying town. When Daniel Defoe visited in 1724, he was displeased by its "meanly-built" aspect and pleased by the price of lobsters. A modern traveler would undoubtedly reverse the criticism.

Dedham

Dedham is a Saxon name derived from settlers named Dydd or Dydda, whose "ham" was a clearing by the ford at a time when the River Stour was still unbridged. Other names can be less reliable. Castle House, for instance, was a clothier's residence in a town which never had a castle. Now open to the public, the house displays the works of its former owner, the twentieth-century painter and sculptor, Sir Alfred Munnings.

But it is in no way misleading to apply all of the synonyms of "charm" to Dedham, one of the best-looking villages in East Anglia. Essentially a one-street village, with Mill Lane branching off toward the mill and the River Stour, its High Street is lined with ancient and delightful buildings that serve as proof that Dedham was an important and prosperous wool center from

Sherman's, High Street

the fifteenth to the seventeenth centuries. The Marlborough Head Inn of about 1500 is on the Mill Lane corner of the High Street. The Sun Inn is another early sixteenth-century, timber-framed building on the High Street, with carriageway and picturesque stable yard.

Among the early Georgian brick houses which line the High Street, Sherman's, just opposite the parish church, implies a connection with the wool trade in its name, formerly Shearman. The Grammar School, in the square adjoining the church, is another fine brick building dating from the early eighteenth century.

Southfields

Dating to 1500 is a particularly interesting group of timber-framed dwellings known as Southfields and located about three hundred yards south of the High Street, past the church and the playing fields. This clothier's house consists of the residence of the master weaver as well as the living quarters of his workmen, offices, and warehouse, arranged around an interior courtyard.

But the attraction of Dedham, and indeed of the entire area, is the parish church of St. Mary the Virgin begun in the year in which Columbus discovered America. It was a time of prosperity when Dedham was becoming an important textile center specializing in bays and says, the finer fabrics in contrast with coarser English cloth. A new church replaced the earlier Saxon or Norman building. Built in the Perpendicular style, with a tower 131 feet high, and situated with its 170-foot-long side to the High Street, St. Mary's is one of the outstanding churches of East Anglia. Its immense size and its attractive and characteristic use of flint reflect the prosperity brought about by the weaving industry.

The long nave of St. Mary's has a pleasantly decorated ceiling and clerestory. Among boss ornaments in the nave ceiling, at the point of intersection of roof beams, is a shield of Dedham in Massachusetts and one of the Commonwealth of Massachusetts. Many left from Dedham in England to help found the colony of Massachusetts Bay, and reminders of the connection between the two Dedhams recur.

The Sherman window in the north aisle is of special American interest. The initials "E.S." belong to an ancestor of General W.T. Sherman of American Civil War fame, Edmund Sherman. He is buried in the churchyard near this window, while the house in which he lived stands across the street. A first cousin of "E.S."—Samuel Sherman—emigrated to Massachusetts in 1634 and settled with his family in a place called Contentment, later renamed Dedham.

In the Lady Chapel, in addition to the fourteenth-century piscina, is a pew honoring the people of Dedham, Massachusetts, who contributed generously in 1967 toward restoration of this church. The bench end shows the seal of the Republic—thirteen stars and thirteen arrows in eagle's claw, signifying the thirteen original states. Carved panels on the back of the pew are decorated with various relationships between the two Dedhams including a design of two intertwined D's, the first house of worship in Massachusetts, and the Mayflower.

American associations are brought into the modern age by an inscription on the first pew of the north side commemorating an event which took place on the twentieth of July, 1969—the first landing on the moon.

Wealthy clothiers had built the magnificent church, and leading citizens established a lectureship in 1578 for sermons resounding with religious zeal. John Rogers, who had acquired a reputation for his pulpit outpourings as Vicar of Haverhill, was called to the Dedham Lectureship in 1605. His eloquence drew huge audiences and made the Tuesday Lecture at Dedham a county event. "Roaring Rogers" died in 1636 and was buried in the parish church. His monument is against the north wall of the chancel, a half figure within a canopy, looking very much like the Shakespeare monument in Stratford.

Dedham suffered from the depression of the 1640s due to a decline in the wool trade, exacerbated by the Civil War and the plague. It never recovered its former position as a thriving industrial community. Gone is the wealth from wool, but it remains rich in beauty.

Its beauty has been captured in the paintings of John Constable, for Dedham is in Constable country. The master was born in 1776 in nearby East Bergholt on the Suffolk side of the Stour. Dedham is in Essex just two hundred yards from the Suffolk border. The scenes on the banks of the Stour, Constable said, made him a painter. The Dedham church tower is a favorite subject and can be seen in his *Dedham Mill*, *Vale of Dedham*, and *A View of the Stour*. So impressed was he with the tower, that he painted it in even where it did not belong. *The Cornfield*, in the National Gallery, portrays the lane leading from East Bergholt across the meadows to Dedham, but the church seen in the distance is artistic license.

In 1821, he painted his masterpiece, *The Haywain*, also in the National Gallery. In that year, he wrote a letter to his wife from the Old Mill House in Mill Lane where he was staying with his sister, expressing the wish "that we had a small house here." So might any visitor well express the same wish. For although Dedham has lost its former prosperity, it has retained something more meaningful: it has inherited a rich abundance of splendid old buildings and a setting which continues to be a source of inspiration and natural wealth.

Denver

It is totally unlike the celebrated city of Denver in Colorado. It is a small village, unknown, sparsely populated, and some would add, unremarkable. Not mountains, but level farmland characterizes a landscape so flat that the cathedral of Ely, some eighteen miles to the south, rises like a gigantic edifice to dominate a vast expanse of countryside all around. And about the same distance to the north lies the sea.

Not even located on a main highway to a major city or grand tourist attraction, its visitors are probably those who have found it by purposeful planning. What is it that they have found? Something actually quite remarkable. Not a Disney-like fantasy world of the future, but a plunge into the peaceful past, an escape from traffic and turmoil, congestion and computers. The other Denver is a pleasant English village that is reminiscent of the great capital city of Colorado in name only.

Denver is located on the road to King's Lynn, a town which prospered from the trade of the Middle Ages, particularly the export of wool, when it was a primary port of the land. Its former wealth is reflected in medieval buildings and architectural gems which make it a delightful town today. Economic history is recalled also in the windswept Norfolk scene, which reveals a landscape still patterned with sheep grazing everywhere.

The fens of isolated Norfolk, past Cambridge and Ely, may bring to mind that quashing line from a Noel Coward play: "Very flat, Norfolk." Yes, flat, for the fens were once desolate and watery stretches, navigable in places and dotted by islands. Indeed, Ely's name means Eel Island.

Settlements took place on the higher patches of land above water level in the days before the great draining turned the unfriendly fenland into fertile farmland. Now the area of largely reclaimed land is characterized by rich agricultural fields yielding crops of wheat and barley, potatoes, sugar beets, onions, celery, carrots, parsnips.

In the midst of all this is Denver, a lovely little village with the usual lovely-little-village elements. The nicely-proportioned parish church at the center is surrounded by leaning headstones in its churchyard to please the eye and suggest age. Although St. Mary's was restored in 1871 with the addition of a north aisle and handsome oak roof in the nave, it goes back to a much earlier date. Most of the windows are fifteenth century, but the East window belongs to the fourteenth. The satisfying interior features a striking barrel ceiling in the chancel, painted blue with red accents, and various memorials, organ, decorated font, and an open tower at the west end through which can be seen ropes for bell ringing.

The embattled church tower lost its spire long ago, in the great gale of 1895. A photograph on the inside north wall shows the church before the storm, crowned by a gold cockerel weathervane. The cock, a dent in its side incurred when the spire toppled, now sits directly on the tower.

Victorians installed a clock on the tower face, possibly as a reminder to frequenters of The Bell, a pretty pub just opposite. Near the pub stands a stone cross engraved with the names of World War I heroes. The memorial, as is always the case in tiny English villages, overwhelms by the disproportionate and unexpectedly large numbers of casualties of the Great War. Near the monument, a public telephone of the unattractive modern variety has replaced the conspicuous, red telephone box which is fast becoming a museum piece, indicating that British Telecom at least has found its way to this sleepy village.

In the center of the village a signpost points the way south to Ely, north to Downham Market, and westward to Denver Sluice (2 miles) and to Denver Windmill (1/2 mile). Visitors to this rural backwater find the sluice and windmill of considerable interest.

The tower windmill at Denver, the finest for miles around, was built in 1835. The grinding mill ceased to operate by wind power in 1941 when it was converted to diesel power. It has been restored and is now maintained by the Norfolk County Council.

It is believed that mills, a familiar part of the Norfolk scene for centuries, may have originated when returning Crusaders brought the idea back from the Middle East. But coupled with the flat reclaimed land in an area of dykes and gabled houses, windmills take on Dutch associations which are compounded further along the road in the sluice that gives Denver a modicum of fame.

Charles I, wishing to make the area habitable and agriculturally prosperous, employed in 1652 a great Dutch engineer; Cornelius Vermuyden undertook the work of draining the marshy fenland and controlling flood waters coming down the River Ouse. He installed a series of long straight dykes, known locally as drains, with spaces in between the dykes forming receptacles for inevitable floods. At the northern end, at Denver, he built a sluice (a channel with a gate) to control the flow and hold back the sea.

The old sluice was destroyed several times, initially by fensmen who, dependent on wildfowl and eels for their livelihood, opposed drainage of their hunting and fishing grounds. Rebuilt, it was again destroyed in 1713, this time by a high flood.

The sluice and lock designed and constructed in 1832 by Sir John Rennie, builder of the Old Waterloo Bridge in London, still stands. Improvements were carried out and its load lightened by the New Denver Sluice of 1959.

The area can offer a good day's education as well as leisure. Yachts, dinghies, and sailboats abound, and boats are available for hire. Fishermen add to the holiday feeling. An attractive riverside pub, Jenyns Arms, is the main gathering place. It offers friendship and refreshment and meals from its position at the center of a complicated junction where sluice and seven water channels—ancient or modern, tidal or for drainage only—combine to prevent flooding. The setting illustrates the deep influence of the sea on the entire area.

Even the local hero has connections with the sea. Captain George Manby, born in 1765 in Denver Hall, achieved fame as the inventor of a rocket-fired apparatus for saving lives at sea. Modified and improved, the device is still in use today. But perhaps a better known figure who emanates from Denver is a fictional one—Lord Peter Wimsey, second son of the late Duke of Denver.

When Dorothy L. Sayers, the popular and respected writer of mystery novels, needed a hero-detective to solve crimes, she could have done no better than to create a character from the East Anglian fenlands and villages she knew so well, having lived in the area for some twenty years. She varied the sibilant sound of the River Wissey (a tributary of the Ouse) to make it suit her unpredictable hero, a man of whims, and gave "Wimsey" roots in Denver where lived a distant relative of the Sayers family. *The Nine Tailors* conveyed the background and mood of the fenlands—rain, floods, rector, and church complete with bell ringing—and used the disaster of 1713 as the basis for the fens flood which occurs in that novel.

While the actual Denver boasts no great house of the aristocracy, it would be lovely to think that Oxburgh Hall, a mere eight miles to the east, is the prototype for Lord Peter's ancestral home. The medieval moated manor house, built in 1482, had been in the ownership of one family for over five centuries; rescued from demolition, the house is now in the care of the National Trust and open to the public, who may well imagine all the accoutrements—unused library. . . fox hunts. . . dinner parties—of such an ancient domain as characterized the Duke of Denver's vast and fictional estate.

Worthy Denver, situated eighty-six miles north of London off the A10 trunk road, is easy enough to reach and certainly worth reaching. Both village and vicinity have much to offer. Downham Market, the business center of the district, is a pleasing town just one mile north, and villages all about beckon with their intriguing old names—Barton Bendish, Shouldham Thorpe, Crimplesham—threatening to immerse the visitor in fenland history for a long and profitable time to come.

This history of Denver is ancient and goes back to the Domesday Book of 1086 when it was called Danefella, meaning "seated in a valley by a water." The name gradually became corrupted over the centuries, with Walter de Denvere mentioned as lord of the manor in 1257. Cited as Denever in the fourteenth century, when John de Denver was the noble lord, the name finally took hold in its present form.

Denver may not have been significant enough to confer its name on an American city, but it has bestowed it, through the centuries, on fictional and actual aristocracy, as well as on commoners. Thus a link was created between the two Denvers. When Denver, Colorado, was founded in 1858, it was named for the Kansas Territorial Governor, James William Denver. Whatever is in a name, that quiet little place in England had it first to bequeath.

Dorchester

So realistically did Thomas Hardy paint the Dorset landscape in his novels, that it is not amiss for the tourist, intrigued by Hardy's treatment, to arrive in Dorchester and make the self-congratulatory observation, "I'm here at last—in Casterbridge." That was the name chosen to represent Dorchester in his novels. Hardy also emphasized the dominant feeling of Dorchester when he wrote in *The Mayor of Casterbridge*:

> It announced old Rome in every street, alley and precinct. It looked Roman, bespoke the art of Rome and concealed the dead men of Rome. It was impossible to dig more than a foot or two deep about the town's fields and gardens without coming upon some tall soldier or other of the Empire who had lain in his silent unobtrusive rest for the space of fifteen hundred years.

After the Roman invasion in the year 43, the town flourished and took on characteristics that still impart to Dorchester that special ancient flavor. Excavations go on continuously, and Roman graves and relics are uncovered frequently. One famous hoard was found in 1936 on the site of the equally famous Marks and Spencer department store in South Street. A Roman necropolis is near Fordington Hill. Other large ones are near South Walks and Salisbury Walks and at Poundisbury. Often, it is just the Roman idea or feeling that exists under or through the modern overlay. For instance, the thriving Roman town had a main highway (Via Iceniana) with streets branching off from it and houses placed in the rectangular blocks. The public buildings, forum, and temple were in the center, around the area of St. Peter's. A careful study of modern Dorchester can reveal the square outline by which the Romans laid out the town in typical grid pattern.

71

Ruins abound. Existing fragments of Roman walls, twenty-eight feet long, are on the east side of Albert Road. Part of the Roman aqueduct which carried the water supply from Maiden Newton (Chalk Newton in *Tess of the D'Urbervilles*) to Dorchester may also still be seen.

For further traces of Old Rome one might visit the fine Dorset County Museum where, amidst the archeological finds, is one skeleton with a Roman dart embedded in the spine.

But evidence of the presence of man in the area goes well beyond the Roman era, some four thousand years back to Neolithic times. Again, the museum traces the history of the area from the Stone Age to the twentieth century. But even prehistoric remains are all around and not limited to the confines of a museum.

Maumbury Rings is a sight which scans the centuries. Originally a sacred circle of the late Stone Age, it was converted by the Romans to an amphitheatre in which gladiatorial combats could be seen by a crowd of thirteen thousand. In the Middle Ages it was used for bull- and bear-baiting and for cock-fighting. Then it became an execution site. In modern times, it is a remarkable memorial to the distant past.

Maiden Castle is a gigantic prehistoric hill fort. It was Mai Dun, the fortress by the plain. Situated two miles southwest of Dorchester, and built originally as a defense, this now-dead city is generally accepted as the finest example in the world of a hill fortress. It occupies 120 acres, extends one thousand yards, and has a width of five hundred yards.

Another spectacular prehistoric sight is the Cerne Giant, a figure cut into the chalk hill overlooking the beautiful little town of Cerne Abbas, eight miles from Dorchester. An abbey was founded in the town (as its name implies) by St. Augustine in 603. But it was destroyed at the time of the Dissolution, and present monastic remains include only the guest house, tithe barn, and gate house with beautiful two-storied oriel window over the doorway. However, the far more ancient giant, a huge primitive man holding a club, is probably of late Neolithic date. He stands 186 feet high, covers one acre of ground, and is the subject of great conjectures by great authorities. He is undoubtedly connected with pagan rites and may be a fertility god, but his secret is lost in antiquity.

The long barrows built by people of the late Stone Age for burial of the dead are plentiful in Dorset.

Dorchester was the tribal capital of the Durotriges when the Romans invaded. When they gave the site of modern Dorchester the Latinized name of "Durnovaria," they retained the root "Durno," meaning "a fist" in the ancient language of the Britons. The somewhat puzzling etymology of the name is interpreted by scholars as possibly suggesting the shape of the landscape with its hill fort, which may have looked like a fist. The Saxon settlers added *ceaster* to the Romano-British root, implying that they found it to be the site of a Roman camp or station. Cities as well as their names evolve, and this one turns up in Domesday Book as "Dorecestre."

Through the centuries, the town had its share of plagues, wars, and other disasters. The Civil War must have made the history of Dorchester in the seventeenth century particularly grim. Then, after the Monmouth Rebellion, at the Bloody Assize of 1685, Judge Jeffreys was sent to Dorchester to punish the rebels. Of three hundred prisoners who were tried, all but eight were sentenced to death. The infamous judge lodged in a house in High West Street, now a restaurant. Naturally, stories teem of Dorchester people hearing the ghost of Jeffreys furiously stomping about in those lodgings.

Hardly had the town recovered from the serious plague of 1595 which killed a substantial portion of the population, when fire burned a large part of the town in 1613, including two of its three churches. Fire, an ever-present threat, again damaged the town in 1622 and in 1775. But the Church of St. Peter, with links to the Roman past and the American future, survived. Built on a site believed to be a Roman temple, the church has relevance to early American history and is a bright spot of seventeenth-century Dorchester history, certainly for Americans.

It was the Reverend John White, rector of St. Peter's, who promoted colonization in Massachusetts. The "Patriarch of Dorchester" died in 1648 and was buried in the church. The last

Church of St. Peter

line of his epitaph reads: "He greatly set forward the Emigration to the Massachusetts Bay Colony, where his name lives in unfading remembrance."

Although he did not himself join any expedition to the New World, it was through his efforts and encouragement that the colony in New England thrived. He was initially motivated by a concern for English fishermen who usually sailed from the nearby port of Weymouth to American waters. How could they receive religious instruction during the nine- or ten-month period when they were away? A resident minister and colony would eliminate the problem.

There were several unsuccessful expeditions under the auspices of the Dorchester Company, but Reverend White persevered. He raised money and ultimately procured the charter that was to be the start of the Commonwealth of Massachusetts. In 1629, with the further support of a shipload of three hundred emigrants, the new colony was firmly established. The colonists landed in an area which they named, naturally enough, "Dorchester" to honor their revered friend. A plaque in the First Parish Church of Dorchester, Massachusetts, reads: "Dorchester Named from the Town of Dorchester in Dorset England." Tablets on either side of the doorway pay further tribute to the Reverend John White.

But it was Thomas Hardy who has probably done more to bring visitors to Dorset than any Chamber of Commerce could hope to accomplish. The area of scenic beauty surrounding Dorchester has been immortalized in his works. There are panoramic views from Blagdon Hill where there is also a monument to the memory of Admiral Hardy, who fought at the Battle of Trafalgar. Nelson died in his arms, addressing his last words to this distant kinsman of the novelist.

Thomas Hardy's inadvertent advertisements are everywhere. Hangman's Cottage, believed to be the actual residence of the public hangman, with its pretty thatched roof, has a charm which belies its function. The popular tourist attraction appears in his story, "The Withered Arm."

Abbotsbury, nine miles to the southwest, was founded as a Benedictine Abbey. It has a well-preserved fifteenth-century tithe barn (mentioned in *Far from the Madding Crowd)*, sub-tropical gardens with a fine collection of exotic trees, and the unusual Abbotsbury Swannery, established by monks in the thirteenth century to supply food for their table. The wildlife nature reserve where swans live and breed is open to the public.

Eleven miles from Dorchester, in the church of Bere Regis ("Kingsbere" in Hardy's Wessex), is the Turberville window. (Yes, *Tess of the D'Urbervilles*.)

It is hardly possible to cover the settings alluded to in Hardy's stories. Just as his memorializations are everywhere, so Dorchester has memorials to him. One statue at the top of High West Street is a life-size representation of the pensive poet resting with hat on crossed knees. Another Hardy memorial statue, at the cottage in Higher Bockhampton where he was born, is the gift of American admirers.

Hardy lies buried in Westminster Abbey, but his heart is in Stinsford, in the "Mellstock" of his novels; there, just a mile and a half from Dorchester, it lies buried under the great yew tree in the churchyard.

The Dorset County Museum in Dorchester may have the finest tribute of all to the literary giant of Dorset. The Thomas Hardy Memorial Room is a recreation of his study in the house he designed for himself at Max Gate. The room includes such memorabilia as his pens and copybook and original manuscript of *The Mayor of Casterbridge*. A roller calendar on his desk is set with the date of his original meeting with his first wife Emma—March 7, 1870.

Hardy's love of Dorset extended to archeology. In a paper entitled *Some Romano-British Relics Found at Max Gate* and read in 1844, always the accurate observer, he recorded some remains found at Max Gate in Fordington Field and called for the reconstructing of bits of evidence of Roman life into a whole, much as has been done at Pompeii. He ponders what Romano-British life was like and asks probing questions: "What social character had the streets, what were the customary noises. . .Did they ever form a busy throng such as we now see on a market day?. . ." He was searching for life among the dead

Thomas Hardy's Cottage in Higher Bockhampton

and dry bowls, bottles, urns, and graves, trying to give life to an ancient world, just as his imagination created life in the pages of his books.

It is much easier, and alas disheartening, to assess life in today's Dorchester, where the customary sounds are traffic noises and even the shouting of voices necessarily raised to overcome the din of vehicles.

Although Dorchester lies in a predominantly agricultural area, it nevertheless has an assortment of inconspicuous industries hidden in the landscape—a printing plant, farm machinery factories, and leatherworks. But the Dorchester Brewery, the largest, has been brewing beer in the town since the seventeenth century. Actually, Dorchester ale has been famous since Tudor times. Hardy pays tribute to it in *The Trumpet-Major* when he says that "it was the most beautiful colour that the eye of an artist in beer could desire; full in body, yet brisk as a volcano; piquant, yet without a twang; luminous as an autumn sunset; free from streakiness of taste; but, finally, rather heady." A real treat is to partake of the local brew at the Smith's Arms in Godmanstone (five miles to the north), an inn which claims to be the smallest licensed house in England.

But the scene in the center of town imparts no feeling of ancient charm, at least not superficially. It suggests rather that Hardy's description of the city is hardly accurate. The dominant feeling is that of a busy, commercial, trafficked center. Instead of Roman appearance, it has roaming crowds of shoppers. Instead of heading for an ancient forum, they throng to a modern shopping arcade. The word "antique" would seem to be relegated to shops which sport that epithet but which disappointingly purvey twentieth-century relics. And all feeling of the past would seem to be relegated to the museum.

Nevertheless, things could be worse—much worse. The gift shops do *not* sell plastic replicas of the Hardy statue or T-shirts with the poet's picture. And the weekly markets *do* sell such delightful local products as Dorset cheese and Dorset butter. Moreover, the county town is not only a convenient center for innumerable not-to-be-missed places within easy hiking or driving distance but can also be a pleasant town which offers escape from the bustle in its tree-lined avenues and river bank, within easy walking distance from the center. Its active character leaves one with the final impression that Dorchester is by no means dormant.

Durham

The visitor arriving at Durham by train is treated to one of the most spectacular scenic views anywhere. The railway is raised on a viaduct above the city, affording a panoramic vista, over rooftops, of castle and cathedral. The motorist too, having made the 260-mile journey from London, may savor the world-famous view across the River Wear which takes in the three great towers of the glorious cathedral. The superb castle and cathedral tower over the wooded banks of the river and make Durham City the prized focal point of County Durham.

It all began around the year 875 when monks from Lindisfarne fled inland to escape from invading Danes, taking the coffin of St. Cuthbert with them. They eventually settled in Durham in the year 995, having chosen for their abbey a naturally defensive position on a rocky peninsula formed by a hairpin meander of the River Wear, which almost entirely encloses the site. The stone church they built was demolished after the Norman Conquest, when Benedictine monks began the building of a new cathedral.

The Normans confirmed the excellent defensive qualities of the site by selecting it to protect the large area of northern England that stretched to the Scottish border. In 1072 they built Durham Castle on the narrow and vulnerable northern neck of the peninsula to guard the approach to the city; and they established a palatinate, or semi-autonomous region, making the castle the residence and seat of power from which bishops of Durham reigned as princes until 1836. The enormous powers of the prince-bishops included the right to levy taxes and grant pardons and extended to administrating their own mint, exchequer, parliament, judiciary, and army.

The city developed around the cathedral and the nearby castle. In the twelfth century, the walls surrounding the peninsula were strengthened, and an inner wall was built to connect castle keep with cathedral choir. Parts of that wall are still visible.

Of the Norman castle, little remains today apart from the chapel and Great Hall. The last prince-bishop gave up the castle in 1832 and established University College. Four years later, Durham University was founded, England's third university and the oldest after Oxford and Cambridge. The castle is today part of the university, which has expanded to possess most of the properties of the Bailey and Palace Green, the area between cathedral and castle.

The Castle Gateway

Through the Norman archway is the Great Hall, still the dining hall of the college. The Black Staircase leads to Bishop Tunstal's Gallery, formerly the bishop's state rooms and now the senior common room of the college. The bishop's chapel at the end of the gallery has become the college chapel. A spiral staircase leads down to the Norman Chapel, the oldest remaining part and a gem of Durham Castle. The city continues to center around this area with its ancient street pattern.

The prime attraction of Durham, its cathedral, was begun in 1093. It is the greatest Norman church in England, with an equally impressive interior featuring huge Norman nave with massive columns, clerestory with windows in arched recesses, and ribbed vaulting, the oldest in Europe, sweeping toward the crossing and the chancel to the Perpendicular east end. The delicate Neville Screen behind the high altar, carved of stone in 1375, serves as a backing for the tomb of St. Cuthbert.

The Nine Altars Chapel, added in the thirteenth century, fills the east end where a stained-glass rose window of 1795 replaced the original. At the west end, the Galilee Chapel of 1175, with slender columns and round arches adorned with zigzag carving, contains twelfth-century wall paintings and the simple tomb of the Venerable Bede, who died in 735; his body was brought from Jarrow to Durham in 1022.

But the monster bronze sanctuary knocker on the front door, the use of which guaranteed outcasts protection within, is merely a replica made in 1980; the original twelfth-century knocker is displayed in the Cathedral Treasury, which also preserves among its acquisitions the seventh-century wooden coffin of Saint Cuthbert and eighth-century illuminated manuscripts. Cloisters on the south side display further treasures, while Palace Green in front of the cathedral leads to the castle.

In the shadow of the castle, at the northern end of Durham, the Market Place evolved and became the focal point of several roads of access to the city, two by bridge. Elvet Bridge on the east (leading to Elvet with its Iron Age fort known as Maiden Castle) was built around 1160 and widened in 1804; Framwelgate Bridge, rebuilt in the fifteenth century, spans the River Wear on the west.

A Saturday market still thrives in the medieval Market Place. But as seen today, the Market Place, the fulcrum of the city, is essentially Victorian. It features the prominent equestrian statue erected in 1861 of the 3rd Marquess of Londonderry. On the north side is St. Nicholas Church (1858). On the west, the town hall of 1851, with a hammerbeam roof modelled after that in Westminster, incorporates parts of the 1665 Guildhall.

All around the Market Place, narrow streets and cobbled alleys with suggestive names provide a treat for walkers. Silver Street must have been the location of the mint of the Bishops of Durham. Fleshergate was the butchers' quarter, and the area between Silver Street and Saddler Street was known as Pullein— that is, poultry.

Saddler Street becomes North Bailey and then South Bailey as it continues toward Palace Green and the city core. University buildings of distinction make it a most attractive street. In particular, Bowes House at No. 4 South Bailey, now part of St. John's College, is a fine sixteenth-century house. Next to it at No. 3 is Sir Robert Eden's eighteenth-century town house, possibly the finest in the Bailey. Also notable are St. Chad's College and the church of St. Mary-le-Bow next to it. Hatfield College was originally the Red Lion Hotel. Attractive houses reveal Georgian facades and conceal older houses as this most delightful street stretches around to the river and to Prebends' Bridge (1778) at the southwest of the river loop. The Wear offers an opportunity for viewing the city by boat or from pleasant footpaths on either side.

The old fifteenth-century corn mill near Prebends' Bridge enhances an altogether pleasant setting which takes in the old Fulling Mill on the opposite bank with perhaps the finest view of the mighty cathedral, a view which has inspired artists. This vantage point gave Turner his scene of Durham with its air of mystery. And it gave Sir Walter Scott those lines engraved at the west end of the bridge:

> Grey towers of Durham—
> Yet well I love thy mixed and massive piles—
> Half Church of God; half Castle 'gainst the Scott.

Certainly, Durham deserves a visit.

It is small wonder that the inspiring name of the English city of Durham was transferred to a new site across the Atlantic. In New Hampshire, the parish known as Oyster River Plantation was incorporated as the town of Durham in 1732. The plantation's first minister, the Reverend Hugh Adams, selected the name as a tribute to Richard Barnes, Bishop of Durham, England, whose administrative plan of government he wished to apply to the new town.

Exeter

Exeter lost its medieval character during World War II when it was severely bombed in retaliation for the allied bombing of Cologne. Afterwards, modern city planners replaced the narrow winding streets with pedestrian zones and modern thoroughfares that are often clogged with standstill traffic. Unlike William of Orange who made a ceremonious entry into the city in 1688 to hear the *Te Deum* sung in the cathedral, the contemporary visitor arriving by car takes the ring road to the new bridge over the river, through a spaghetti junction. Nevertheless, the great cathedral city is blessed with valuable and historical treasures at its core, making it a worthwhile tourist stop.

The present High Street was a ridgeway when the Celtic people, the Dumnonii, settled there between the second and third centuries BC. When Exeter was Isca Dumnoniorum and capital of the entire southwest, the Romans established a street plan that is still discernable. At the central crossroads, North Street becomes South Street; and, at the intersection, Fore Street continues on as High Street as it runs uphill, west to east, from the River Exe toward the cathedral.

The cathedral, its exterior virtually unscathed by bombs, remains the focal point of the city. It sits on a quiet street, nearby but apart from the noise and turmoil of the main thorough-fares, where one may contemplate the magnificent west front covered with carved figures and the two flanking Norman towers.

The interior is outstanding for its long nave with Purbeck marble columns and for the Gothic rib-vaulting which extends for three hundred feet, the longest continuous stretch in the world. An overhanging minstrels' gallery on the north side is carved with painted angels holding medieval instruments. A marvellous Bishop's Throne, sixty feet high, was constructed of oak in 1316 without nails. When the stonework of the east window of 1390 decayed, it was rebuilt in the Perpendicular style with the original glass of 1302. A fifteenth-century clock in the north transept shows the sun and moon revolving around the earth.

Through a fourteenth-century gate-house southeast of the cathedral stands the Bishop's Palace, near the place where Roman soldiers were garrisoned in the first century. Within the cathedral Close, remains of a Roman bath-house and basilica were uncovered after the 1942 bombing raid.

Mol's Coffee House

Buildings around the Close are clerical as well as commercial. The most conspicuous is the Royal Clarence Hotel, built in 1769 with an assembly room running back to the High Street. It became the first English inn to use the French name of 'hotel' when it was run by a Frenchman, Philip Berlon, and known simply as Philip's Hotel. The name change came about after the Duchess of Clarence, later Queen Adelaide, visited in 1827. Other guests have included Lord Nelson and Tsar Nicholas I.

A narrow passageway on the side of the Royal Clarence leads to the Ship Inn, Sir Francis Drake's favorite, where he regaled Raleigh, Frobisher, and Gilbert with sea stories. Also in the Close, near St. Martin's Church, is Mol's Coffee House, a black-and-white building of 1596, with galleon-style bay windows, now just one of a number of high-class shops that fringe the cathedral green; but in the eighteenth century an Italian named Mol served the new beverage to gentlemen who discussed politics over the fashionable drink.

In one house a few paces away, Anthony Trollope often visited his mother's friend, Fanny Bird, the model for Miss Jemima Stanbury in his 1869 novel, *He Knew He Was Right*. "In Exeter the only place for a lady was the Close," pronounced the respectable fictional character.

Exeter is fortunate. Across the way from the two lanes that lead into the Close, in the High Street, the remarkable Guildhall has escaped both the bombs and the builders. Dating from 1330, it is said to be the oldest municipal building in the country. Of particular note is the ornate upper-story portico which rests on a row of granite pillars, built in 1593 when Exeter's wool trade was at its peak and merchants were among the wealthiest in the land. Still used as a council chamber, the building is open to the public who may view various symbols of the city's history.

The Turk's Head next door was visited by Charles Dickens, who based the character of the fat boy in *The Pickwick Papers* on a boot boy who worked at the inn at the time.

Also of great age is Wynards' Almshouses, built in 1436 by a wealthy wool merchant to house twelve poor men. Restored, it now serves as an advice center for several voluntary organizations of the city.

The Guildhall

Alongside the Church of St. Mary Steps is Stepcote Hill, with steps and cobbled gutter which served as a main street in medieval times. Two houses past the church, dating to about 1500, have been preserved in situ; but an even earlier house on the opposite side, one of the oldest timber-framed houses in

87

Europe, was moved to the present site from Edmund Street when it was threatened with demolition in 1961.

Believed to be the earliest Norman castle in Britain, Rougemont— meaning "red hill"—was chosen by William the Conqueror in 1068. But only the stone tower and main gateway of the old fortress remain within this northwestern corner of the city. The old Roman wall can be seen in the gardens which now cover the inner bailey.

Still in existence near the West Gate are the remains of the old red sandstone bridge of the twelfth century. Eight of its original eighteen arches stand in gardens near the ruins of St. Edmund's Church. Proximity—only a five-minute walk from the High Street—to the River Exe (a Celtic word for a river abundant with fish) serves to remind the visitor of Exeter's long connection with the sea. The famed maritime museum is a further reminder, as is the old Customs House, built in 1680. Also belonging to the same period is the Quay House, open to the public with displays of Exeter's history.

Against such a background of antiquity, Exeter's airport and its rapidly expanding university help to return the visitor to the reality of modern times.

It is also against the ancient background, that the source of the New Hampshire town of Exeter springs to the fore. First settled in 1638 and known as Squamscott, the new town was managed by a group of English colonizers headed by the Reverend John Wheelwright; the "Exeter Combination," as the colonists were known, borrowed the name of the lovely English town in Devonshire for their new settlement.

Falmouth

Falmouth is a lovely seaport town with a climate so mild that oranges and bananas grow, palm trees line some streets, and tropical plants thrive in public gardens. The scene suggests a Mediterranean seaport town.

But Falmouth is a Cornish seaport town located on the English Channel, where its strategic importance has been appreciated through history. Sir Walter Raleigh called attention to the possibilities of the Fal Estuary as an important deep water harbor. Earlier, in the 1540s, Henry VIII had built Pendennis Castle as part of his scheme for protecting England from possible attack by his Roman Catholic enemies. Built of Cornish granite, the defensive castle stands at the entrance of the Fal, guarding, together with its twin castle at St. Mawes on the other side of the river, the inland towns. These are perhaps the best preserved Tudor castles in the land—with apartments, huge fireplaces, spy holes, circular central towers, and angled bastions.

Although its history as a trading port is quite ancient, the town of Falmouth itself goes back merely to the seventeenth century. Prior to that time, the site contained the small fishing hamlet known as Smithwick, or more popularly, Penny-come-quick. (Actually, "Pen-y-cum-cuic" meant "head of the creek.")

Also in the district was Arwenack, the major house of the Killigrew family—gentlemen, pirates, and privateers—who are credited as the founders of Falmouth. They started the building of the town in the 1620s, and Sir Peter Killigrew continued to foster its development after the Restoration in 1660. He secured a charter from Charles II and gave the new and descriptive name of Falmouth to this little place. With its superb situation and nearly ten square miles of sheltered water, Falmouth soon outstripped rival Cornish ports.

The selection of Falmouth as a mail packet station in 1688 insured the prosperity of the town for the next century and a half.

Pendennis Castle

Falmouth became the western packet headquarters because of its proximity to the continent and because of its good harbor. Sailing ships could enter and leave quickly, carrying mail to Spain and Portugal, to the West Indies and the Americas. The packet boats were privately owned and hired out to the post office. From this service came much of the town's wealth, still in evidence. For example, Captain John Bull, who carried mail during the Napoleonic Wars on a famous ship called the *Duke of Marlborough*, built a mansion at Swanpool that he named after that ship. The Customs House on the quay also dates from those times.

The contemporary Poet Laureate, Robert Southey, wrote in 1802 a vivid account of Falmouth life in its prosperous heyday:

> ...Everybody is in a hurry here; either they are going off in Packets, and are hastening their preparations to embark; or they have just arrived, and are impatient to be on the road homeward. Every now and then a carriage rattles up to the door with a rapidity which makes the very house shake. The man who cleans the boots is running in one direction, the barber with his powder-bag in another; here goes the barber's boy with his hot water and razors; there comes the clean linen with the washerwoman; and the hall is full of porters and sailors bringing in luggage, or bearing it away; now you hear a horn blow because the post is coming in, and in the middle of the night you are awakened by another because it is going out. Nothing is done in England without a noise, and yet noise is the only thing they forget in the bill!

Falmouth lost the packet service (and perhaps the noise) in 1852 with the coming of the railway. Nevertheless, because of its continued use as a port, it did not suffer a serious decline. More important, its possibilities as a seaside resort were developed, and it became a vital ship repair base with huge dry docks. It was used in the Second World War, in June 1944, as a base for the dispatching of ships and troops for the D-Day invasion.

Falmouth is an active commercial and industrial town but with enough antiquities and memories to conjure up an image of the past. The waterfront is still the heart of Falmouth. Even the main shopping street curves to follow the shore line and is connected to quays by alleys called "opes." Old combines with new. A fragment of the Killigrew mansion remains as part of a modern house on the sea front—a reminder of the ambitious beginnings of a town still thriving in a state of warm, friendly, and handsome youth, a town which faces the sea as if to pay tribute to the source of its vitality.

Falmouth is as "young" as some American cities, having recently celebrated its tercentenary. In 1961, just three hundred years after receiving a charter from Charles II, Falmouth was granted a coat of arms which includes in its heraldic design a packet boat in full sail at its crest. At the base is water, representing the Fal and the sea, as well as a most appropriate motto, "Remember."

Framlingham

Compared with its counterpart in England, Framingham in Massachusetts has lost its "l"—and more. Although early colonial records, including the Diary of Cotton Mather, spell the American Framingham with the "l," it began to be common practice some three hundred years ago to omit the letter in writing. In speech, the letter is frequently unsounded, even today.

The approach to Framingham in New England is via a busy highway and major shopping centers. The visitor to Framlingham in old England is confronted with the sight of a noble ruin—a huge castle which rises high up on a mound built and used by the Saxons as fortification some twelve centuries ago in "the village of Framela's people."

The castle is the focal point of Framlingham, but Framingham history focuses on one man—Nicholas Danforth. A churchwarden in the reign of King Charles I and an early emigrant to the New World, this seventeenth-century Framlingham yeoman owned seventy acres of land and lived in a timber-framed house of two floors believed to have been built no later than the reign of Queen Elizabeth I. He was probably born in that house, now called New Street Farm and now a private residence located about a mile and a half from the town center on the road to Saxtead.

Six years after his wife died, Nicholas Danforth left for America with six of his seven children. That was in the year 1635, just fifteen years after the *Mayflower* had sailed. He died in Cambridge, Massachusetts, in 1638, aged about forty-eight, but left behind a progeny of notables said to be represented in every state of the United States.

His eldest son Thomas held several political and civic offices including Treasurer of Harvard University, Deputy Governor, and Associate Judge of Superior Court. His second son, Rev. Samuel Danforth, was one of the first five fellows of Harvard.

One descendant, Josiah Quincy, became president of Harvard University. Another descendant, James Abram Garfield, became president of the United States.

It was the eldest son of the founder of this distinguished family who founded the American town of Framingham. Thomas Danforth owned the plantation known originally as Danforth's Farms before it was named for the original and beloved home of the first inhabitants from the charming market town in Suffolk.

In that market town today, the castle is the main attraction. An earlier castle on the site is believed to have been built by the Saxon kings of East Anglia, the last of whom was Saint Edmund. He was captured by Danish invaders and died at their hands in 870, a Christian martyr. Framlingham became the seat of the earls and dukes of Norfolk, and the second earl of Norfolk, Roger le Bigod, built the castle as seen today, circa 1190. This powerful nobleman, present at the coronations of King Richard I, King Richard II, and King John, appears as Lord Bigot in Shakespeare's *King John*.

Framlingham Castle

94

The medieval walls of Framlingham Castle are over forty feet high and in places eight feet thick, with foundations that go thirty feet below ground. It is surrounded by a double moat, now dry, with a bowling green between the inner and outer moats. The ancient game of bowls is believed to have been played here some six hundred years ago.

Perhaps the most famous historic event that took place in the castle occurred in the summer of 1553 when Princess Mary withdrew to the Duke of Norfolk's Castle of Framlingham after the death of King Edward VI. With the help of supporters, she was here proclaimed Queen of England.

Framlingham has a fifteenth-century church with a nearly one-hundred-foot-high tower. The chancel of the Church of St. Michael is an unusual architectural feature because of its disproportionate size. As if giving way to Middle Ages spread, it is wider than the nave. It is also full of monuments and history and monumental history.

Chief among the monuments is that of Henry Howard, known in literature as the poet Earl of Surrey. Surrey was convicted on a charge of high treason and beheaded on Tower Hill. He lies resplendent alongside his wife, with two sons kneeling at their feet and three daughters at the head of the alabaster tomb. He wears rich red robes trimmed with white ermine and gold clasp. Sadly, his coronet rests by his pillow to signify that he was beheaded.

Another of the six effigy tombs in the chancel is that of Surrey's father, Thomas Howard, third Duke of Norfolk (1473-1554) and also the uncle of Queen Anne Boleyn and Queen Catherine Howard. In the chancel wall above his tomb is the helm which he wore at the Battle of Flodden in 1513. The words on the collar of his effigy, "By the Grace of God I am what I am" refer to his fortuitous escape with head intact when Henry VIII conveniently died on the day preceding the scheduled execution. In one of those ironic twists of history, if Henry VIII had had the grace or judgment to die one day sooner, the Duke's son, the poet Surrey, would have avoided his untimely death as well.

95

Relatively recent history of the town includes the new Framlingham College which was founded as the Suffolk memorial to Prince Albert. Its facts and features include an eight-foot bronze statue of the Prince Consort, an enrollment of over four hundred boys, over thirty-five acres of playing fields, and an impressive site on high ground with views of the castle and church. Opened in 1865 by Queen Victoria, this "new" bit of Framlingham history has already celebrated its centenary.

The small and appealing town is filled with a sense of history, peace, and beauty. In the center, in the attractive Market Hill, is the Crown Hotel, which dates in part from the sixteenth century. Its massive open timbers of old English domestic architecture and eighteenth-century staircase with carved balustrade are admirable.

Also in Market Hill is the former Guildhall, now called London House, with an interesting oak-paneled room. It has been a drapery store since it was rebuilt by Simon Pulham around 1584.

New Road offers a mixture. Sir Robert Hitcham's Almshouses, a picturesque group of six houses for men and six for women was built in 1634 from stones of buildings razed inside the castle walls. A cinema and a factory in New Road are modern elements which combine with good views of castle and church.

The charming country town with first-rate shops, hotels, and pubs offers a retreat just 88 miles from London to a romantic vision summed up by the poet James Bird (1788-1839) in his poem "Framlingham":

> ...fair castled Town,
> Rare spot of beauty, grandeur, and renown,
> Seat of Anglican kings!

Gloucester

That Gloucester is located on water seems natural enough and unremarkable to anyone who knows and loves Gloucester in Massachusetts. Gloucester in England is also located on the water, but on a river, the Severn, which broadens out before flowing into the Bristol Channel to give the appearance of an inland sea. The English Gloucester offers further contrast: Instead of pounding waves against dignified masses of rocks, instead of sandy stretches of beach, it is a thriving inland port and manufacturing city that rises in the center of the Severn Valley.

As a modern industrial center, it is not a particular tourist attraction. Nevertheless, it has great antiquity and provides sights that are worthy of attention. First of all, the tourist approaching the capital of Gloucestershire from the east, from London, passes through exquisite countryside. The famed region of hilly country known as the Cotswolds offers sweeping landscapes with gently rolling hills which unroll and reveal hidden villages. Vistas are greatly enhanced by the characteristic local building stone used for houses, cottages, and churches as well as for the stone walling which separates fields. If visitors can get through this photographer's paradise without being permanently entranced by the typical Cotswold architecture, Saxon or Norman churches, quaint bridges, or Roman remains, they come to the cathedral city of Gloucester.

The medieval city has been allowed to disappear almost entirely. But two monastic ruins remain, the Black Friars and Grey Friars; and the fifteenth-century St. Mary de Crypt and St. Nicholas are just two among many churches to be singled out for rewarding study. Other interesting medieval buildings include the fourteenth-century New Inn with beautiful galleried courtyard and the fifteenth-century almshouses near Westgate Street. The Old Customs House is in Quay Street. And remains of the Roman city wall may excite wonder.

The House of the Tailor of Gloucester appeals to many. The minute shop chosen by Beatrix Potter for her *Tailor of Gloucester* is now a museum depicting the life and art of the famous author, with a reconstruction of the Tailor's kitchen.

But the great attraction of the English cathedral city is, after all, its cathedral. Formerly an abbey church begun in 1089, the cathedral still displays elements of Norman architecture—the crypt, great nave pillars, chapter house. Although the earlier church was largely destroyed by fire in 1122, rebuilding over the fourteenth and fifteenth centuries has resulted in the present building.

The glorious east window, the second largest medieval stained-glass window in the country (after that of York Minster), is a highlight of the cathedral. Its forty-nine figures, with a central theme of the Coronation of the Virgin, allows light through its translucent tints and gives the east end of the choir, which it nearly fills, a magical effect. It may be thought of as one of the first war memorials, for it was installed in 1350 by the Governor of Gloucester Castle to commemorate the Battle of Crecy in 1344.

The superb fan vaulting of the cloisters, dating to 1351, was afterwards copied in Henry VII's chapel at Westminster. On the east side of the cloisters is the chapter house where one of the great events of history took place in 1086 when William the Conqueror ordered the compilation of the Domesday Book, that written record of all English property.

History records also the hasty coronation, upon the sudden death of King John in 1216, of his eldest son Henry, a youth of nine. Henry III was made to pledge at his coronation to govern according to Magna Carta, a pledge he broke, as did his father earlier.

Another significant event occurred just over a hundred years later, in 1327, when Edward II was murdered at Berkeley Castle and buried in the cathedral. Great numbers of pilgrims flocked to the medieval city to visit the tomb of Edward II. Nowadays, crowds flock to the impressive twelfth-century Berkeley Castle, open to the public, and located just sixteen miles southwest of Gloucester on the fringe of the Cotswolds.

Gloucester Port

In 1541, Henry VIII converted the great abbey to a cathedral. But it was Elizabeth I who declared Gloucester a Port City and granted a charter in 1580 to establish the city docks. Thus was the commercial fate of the city sealed. Excellent facilities for water transport, as well as later rail, road, and air transport, have insured Gloucester's continuing industrial development.

But even before the Norman Conquest, it was active when the casting of iron was a chief industry. And a bell foundry, dating from about 1270, existed for nearly six hundred years. Small manufactures emanated here as well when the city was known for a long time for the production of pins, introduced by John Tylsley in 1626 to replace scraped, trimmed, and dried thorns. As a twentieth-century manufacturing center, its products are as diverse as aviation equipment, man-made fibers, farm equipment, and ice cream.

The modern, commercial city, predictably, bustles with shopping centers, cattle market, parks, hospitals, schools and libraries, museums and art gallery. But a quiet haven for rest and contemplation exists in the cathedral precinct, and further grounds for contemplation may be gleaned at the Gloucester Museum with its archeological finds from prehistoric, Roman, and medieval times, excavated from underneath the modern city.

With roots in antiquity, Gloucester carries a name over two thousand years old. Before Roman times, its name was "Caer Glow," with *Caer* meaning an enclosure or town and *Glowe* generally agreed to mean "fair." After the Roman invasion, it was given the Latinized form of *Glevum*. It reverted, in Saxon times, to something like its original form: Glowcaestre, Glewan-cester, and Glew-ceastre are Saxon versions of Glevum Town or Fair City.

The Beautiful Port, Champlain named the American city—a sobriquet that hearkens back to the old English city. How auspicious that the American settlers aptly renamed their city in remembrance of the one from which they had emigrated!

Great Barrington

The Cotswolds is an area of undefined borders and immaculate beauty. Drystone walls separating sculptured fields, houses of honey-colored stone catching the changes of light, villages sheltering in valleys formed by small rivers—these are a few of the features of the fashionable Cotswolds. Less popular in earlier times, the area was denigrated in Shakespeare's *Richard II* when Northumberland, using words that are no longer applicable, refers to "these high wild hills and rough uneven ways."

Here in this region of gentle hills, not in the Berkshires of Massachusetts, but in the Cotswolds of England, how strange it seems to come upon a little village with a name that belies its size—Great Barrington.

Located in the Windrush Valley, some twenty-three miles west of Oxford and just inside the eastern border of Gloucestershire, the village consists of a double row of pleasant seventeenth- and eighteenth-century stone houses along a street enclosed at one end by a Palladian mansion known as Barrington Park. Built for Earl Talbot, Lord Chancellor in the reign of George II, the prodigious house is set away from the village in a landscaped park with temples and statues. Mostly hidden from view, the private estate may be glimpsed through wrought-iron gates, but a better peek may be obtained from the footpath between Windrush and Little Barrington across the valley. Alongside the property, a driveway leads to a Norman church with an embattled tower containing six bells. The interior of St. Mary's is notable for its high chancel arch, splendid monuments, and intriguing stone corbel heads.

And that is essentially all there is to the entire scene which comprises Great Barrington—dwellings, great manor house, and church. It all conspires to give the appearance of a feudal village through which flows a small river.

But no village would be complete without a friendly pub, and Great Barrington, located on the north side of the River

Windrush, is complete. The road through the village crosses the river and leads to the Fox Inn, with its promise of refreshments and companionship and possibly a game of skittles.

The seventeenth-century stone bridge over the river was built by local mason Thomas Strong. Quarries in the Windrush Valley had been worked from early medieval times, but the Strong family of Great Barrington were the supreme artists in stone in the seventeenth century. Master-owners known for their fine craftsmanship, they were also employed by Sir Christopher Wren in the building of St. Paul's in London.

Beyond the bridge and The Fox, the road continues to the larger town of Burford, just three miles to the east. Again a stone bridge made from local quarries crosses the winding Windrush. The narrow medieval bridge, having outlived the attempts of traffic experts to widen it, leads into the impressive, broad main street of Burford, a town well worth exploring, as are other villages in the Valley such as Taynton and Windrush. And if you think Great Barrington is small, try Little Barrington.

Great Yarmouth

Although Yarmouth—Great Yarmouth, actually—achieved fame primarily as a herring port, it flourished also as the principal seaport on the eastern coast of England. From Great Yarmouth, some pilgrims embarked for Leyden before eventually sailing for the new colonies. Among the first arrivals to the place that was to become the Cape Cod town of Yarmouth were Anthony Thacher and Richard Sears, both from the Pilgrim Leyden Congregation.

While some believe that it is another Yarmouth, the one on England's Isle of Wight, that is the source for the name of the Cape Cod town, the evidence suggests that early emigrants, most of whom were originally from East Anglia, remembered and honored the East Anglian seaport on the Norfolk coast. At their request, the new Massachusetts township was first recorded as Yarmouth in January 1639.

Great Yarmouth has been a thriving port for centuries. Indeed, it is because of its importance as a port that Yarmouth was badly bombed during World War II. Much of the old town has been rebuilt, as was the Parish Church of St. Nicholas, at the northern end of the Market Place. The church, restored to its original medieval form, remains the largest parish church in England.

An attractive array of Georgian houses stands by the church, and in between two of them is the half-timbered house in which Anna Sewell, author of *Black Beauty*, was born. Another literary connection was endowed by Charles Dickens when he set *David Copperfield* in Great Yarmouth.

Opposite the church lies the wide and pleasing Market Place, newly pedestrianized. A lovely building in the northeast corner with a not-so-lovely name, the Hospital for Decayed Fishermen, of 1702, displays statues of Charity and Neptune and underscores the town's connection with the sea. In the opposite corner are a few Rows that survived war bombing, while in Market Row and Broad Row are Victorian-fronted shops.

Rows, or narrow alleys, comprised the ancient core of Yarmouth around which two entirely different areas developed. On the seafront to the east is the late Victorian holiday resort with pier entertainments, amusement arcades, pleasure parks and fun fairs, as well as the Maritime Museum, formerly a home for shipwrecked sailors.

Away from the sandy beaches, on the other side of Great Yarmouth, on the River Yare, elegant old merchants' houses survive on the South Quay and face the still working port. The Elizabethan House at Number 4, despite a deceptive Georgian facade, was built in 1596. Now a museum of domestic life, it displays Elizabethan rooms with oak paneling and intricate plasterwork ceilings. The splendid house at Number 20 was built in the eighteenth century by the wealthy merchant, John Andrews and is now the home of the Port Commissioners. Also attesting to former port prosperity is the Old Merchants House on South Quay at Row lll, a seventeenth-century brick townhouse maintained by English Heritage.

This row leads to Tolhouse Street and the Tolhouse Museum, an early building with an external staircase dating to 1235, which served as a courthouse and jail for centuries and now serves to display objects pertaining to the town's history as a port, fishing center, and holiday resort.

Beyond is Hall Quay and the administrative center with the Town Hall of 1882 and the early seventeenth-century Duke's Head Hotel. And nearby is moored the Lydia Eva, now a floating museum on the history of the herring industry.

Along the quays in the modern and flourishing port, ships from various parts of Europe unload cargoes ranging from grain to cement. It is an increasingly noisy place as wealth in recent times comes from North Sea oil and gas rather than from herrings.

Once one of the largest fishing centers of England, Yarmouth now offers the tourist the opportunity to challenge the superlative assessment made by Peggotty, the Dickens character of *David Copperfield* fame, that Yarmouth was "on the whole, the finest place in the universe."

Groton

A road map of East Anglia shows a complex of red, green, and yellow lines which indicate the primary and secondary roads leading from London through some of the most beautiful scenery of Suffolk. But a road that is merely lined in and without color—a "white" road—leads to the tiny, colorful village of Groton, sixty-three miles northeast of London. Actually, Groton is not even a village, but a parish, located within a cluster of surrounding hamlets.

Although the history of Groton goes back to before Domesday Book, that survey ordered by William the Conqueror in 1086, it is relatively recent history that singles out Groton and locates it on the American Heritage Trail. Groton is a historical shrine, for it was the home of John Winthrop, whose influence in the early history of New England is immeasurable.

The center of every English village is the church; and in Groton, the Church of St. Bartholomew, which hides in quiet countryside, dominates. A prominent sign before the entrance announces:

> John Winthrop, Leader of the Puritan Emigration to New England in 1630 Founder of Boston and First Governor of Massachusetts U.S.A. was Patron of this Church and Lord of the Manor 1618-1630.

Already, at the time of the Norman Conquest, the county of Suffolk was given the nickname of *Seelig* (meaning "Holy") for its great number of parish churches. Some four hundred existed at the time of Domesday, and nearly one hundred more were added by mid-thirteenth century. Because England's wealth came largely from wool, Suffolk, an important wool-weaving center, was prosperous enough to procure the best craftsmanship for its numerous churches. So it comes as no surprise that the Groton Parish Church of St. Bartholomew's, with its nave and

Groton Parish Church

clerestory, chancel and aisles, and its thirteenth-century Early English tower is an impressive building.

All around are memorials and tombs of John Winthrop and his family. The large stained-glass east window of 1875 was erected to the memory of John Winthrop. His father and grandfather are buried in the chancel, as are his first two wives and the infant daughter of the latter. His third wife died in Boston in 1647, and he married for a fourth time in 1648. He himself died in 1649 and is buried in the grounds of King's Chapel in Boston.

An interesting entry in the church register records the marriage in 1622 of Lucy Winthrop, sister of John Winthrop, to Emanuel Downing, later a resident of New England. Their son, Sir George Downing, one of the first graduates of Harvard

College in 1642, is remembered by the famous London street named in his honor.

The windows of the organ chamber in the north aisle contain shields of the Winthrop family and of Massachusetts, with an Indian in gold against a blue background.

Of the ancient manor of Groton, nothing remains but the mulberry tree which stands in what was once Winthrop's garden. Mulberry trees must be signs of greatness, for one of equal renown stands in the court of Christ's College, Cambridge—Milton's mulberry tree.

The manor of Groton belonged to the Abbey of Bury St. Edmunds in the eleventh century. After the Dissolution of the Monasteries by Henry VIII, Adam Winthrop purchased it in 1544 and became the first Lord of the Manor and Patron of the Church, a position later inherited by his grandson. John Winthrop, born in 1588 in the adjoining village of Edwardstone, sold his interest to follow up on new interests in the New World.

Several shiploads of Suffolk families, including many from the Groton district, sailed with John Winthrop in 1630 when he went out on the *Arabella* to the colony of Massachusetts. He proved to be an intelligent and able leader who earned for himself the accolade, "The Father of New England."

The governor's son, Deane Winthrop, gave the name of his native place of Groton in England to the Massachusetts settlement, which was incorporated in 1655.

The origin of the name Groton is derived from the Anglo-Saxon *grotan*, meaning coarsely ground oats, as in the word *groats*. It has been suggested that the tract of land which overlooks the valley of the River Box was gravelly or gritty. If the Old English word *grot* means a particle, it also appropriately names a grain of sand on the map.

But it must be true that great groats from little gravels grow, for the sailing from the Groton district in 1630 transplanted not only names of old Suffolk homes and distinguished men and women, but established a strong and everlasting Anglo-American bond.

Haverhill

The history of modern Haverhill, on the Essex-Suffolk border in England, begins in the eighth century with an Anglo-Saxon settlement on the western edge of the present town. It was a trading community that profited from its position as a half-way stage between Cambridge and Sudbury. The trade probably specialized in oats, as *haver* means oats.

Listed in the Domesday survey of 1086, Haverhill developed in the Middle Ages as a typical East Anglian market town. By the middle of the thirteenth century, all roads entering Haverhill passed through the market place for easy collection of tolls. Wealth was based on the weekly market and on important annual fairs. The town thrived, and a new church was built in the thirteenth century.

Wool and a booming economy came to Haverhill in the fourteenth century. The town rebuilt and enlarged the parish church and established handsome guild and civic buildings. Prosperity continued into the seventeenth century with the arrival of new weaving techniques brought by Flemish refugees. Sales of Haverhill cloth are recorded in London. Gurteen, a weaver from nearby Clare, moved to Haverhill and established the family industry that is still a vital part of the town's economy.

Its counterpart on the banks of the Merrimac River in Massachusetts was settled in about 1640 by the Reverend John Ward, who named it Haverhill for his home in England. Dr. Mather writes of this eminent worthy of New England and states that "his grandfather was that John Ward the worthy minister of Haverhill, and his father was that Nathanial Ward, whose wit made him known to more Englands than one."

Alas, little remains of the homestead he left. A fire of 1665 destroyed the town. Its church was greatly damaged, its guild and civic buildings were utterly demolished, and the population was reduced to unemployment and poverty. The town recovered slowly during the eighteenth century. It was the middle of the nineteenth century, however, with the coming of the Industrial Revolution and the railroads, that was a period of development.

Consequently, most of present-day Haverhill is a Victorian factory town with rows of dreary brick houses. Apart from Weavers Row with its twelve fine weavers' houses, there is little of interest for the tourist. A major landmark is the factory of D. Gurteen & Son. On the edge of town, a sign suggests Victorian delicacy by pointing to the "Civic Amenity"—town dump.

Haverhill, a small town fifty-five miles northeast of London, offers views of the hills and visions of the past.

Hingham

In the lovely Norfolk town of Hingham, elegant eighteenth-century houses cluster around a large green and a smaller one, as if ready to serve as the film setting for a Jane Austen novel. Gentry from the surrounding countryside built their Georgian houses here for the winter assembly and for times when weather conditions would make rural roads impassable. Completely unspoiled, Hingham also exhibits trans-Atlantic links everywhere.

Hingham in Massachusetts, formerly the colony of Bare Cove, was founded in the early seventeenth century by Robert Peck, a rector who left old Hingham for the freer religious climes of the New World, first destroying the church altar rail, which he believed to be idolatrous. He was joined by another Hingham resident whose name is perhaps better known.

Samuel Lincoln, an apprentice weaver baptized in the church in 1622, took ship in 1637 together with other Nonconformist Hingham weavers. While he produced nothing himself in the New World to change the course of history, he did produce a direct descendant, a great-great-great-great grandson, who managed to make a great deal of history. A bronze bust of Abraham Lincoln in the north aisle of St. Andrew's Church is a gift to the English town from the American people.

Also in the church, near the south door, is an ancient baulk of timber taken from the Old Meeting House at Hingham, Massachusetts, said to be the oldest church in the United States in continuous use. But the English St. Andrew's Church, with a particularly fine tower, is a splendid example of the fourteenth-century style of architecture. Among the many treasures contained inside is a fifteenth-century monument, one of the grandest in England, to Lord Morley who died in 1435.

Returning to American ties, a block of granite embedded in the wall of the Post Office was given by Hingham, Massachusetts; the American town exhibits, in turn, a mounting block from Hingham, Norfolk, formerly used outside the church, that bears an inscription announcing that the boulder "was known to the Forefathers before the Migration."

Hingham, located some dozen miles west of Norwich and some one hundred miles miles northeast of London, is a pleasant place to visit, with usual village shops and the friendly and impressive White Hart Inn stretching almost the entire length of the north side of the Market Place. Rebuilt in the 1770s, the White Hart was sited on the coaching road that once contributed to Hingham's prosperity. Nowadays, it would seem most suitable for bonds to be tightened here, with visitors from abroad stopping for respite and local brew from the handpump, as they confirm trans-Atlantic connections.

Ipswich

Ipswich was Gipswich. It was the dwelling of Gipi, the leader of those seafaring Anglo-Saxons who took full advantage of a position at the head of the River Orwell estuary where the River Gipping joins it. The settlement became a flourishing trading community by the ninth century and was one of the most prosperous of English towns at the time of the Norman conquest.

In the year 1200, King John granted the town its first charter, and the seal of the town further corroborates the importance of trading; it depicts a primitive ship, the earliest known example of the type of sailing vessel with a fixed rudder, a type that could easily ply its way up the river along Ipswich quays.

Trading has continued to play a role in the fortunes of the town. In the thirteenth and fourteenth centuries much of the local wool, considered the finest in Europe, was shipped to Flanders to be made into cloth. In the next two centuries, wool was made into cloth in England and then shipped. Only when the industry moved to the west and north from East Anglia, in the seventeenth and eighteenth centuries, did the town decline. But with the opening up of railway communication and with harbor improvements, industry grew and Ipswich revived.

The extensive woolen cloth trade of the Middle Ages has been supplanted by a wide range of cargo. Today, Ipswich manufactures and exports such items as farm machinery, construction equipment, heating systems, garments, baked goods, and beer. And Ipswich remains one of the busiest and most important ports of the east coast.

One variation on a maritime theme links Ipswich with America. In 1614, Captain John Smith landed in a place known as Agawam and wrote that it would make "an excellent habitation, being a good and safe harbor." The region was to be settled by the the son of Governor John Winthrop, who purchased land from the Indians for twenty pounds and led a group of settlers to the site in 1633.

The land was indeed bountiful, and Pastor Higginson of Salem recorded that "our turnips, parsnips and carrots are here both bigger and sweeter than is ordinary to be found in England." And perhaps many would maintain that he was carried away on an air of enthusiasm when he recorded that "a sup of New England aire is better than a whole draught of English ale."

In any case, the settlement took root and thrived, and on 4 August 1634, the court decreed that Agawam be called Ipswich after Ipswich in England "in acknowledgement of the great honor and kindness done to our people who took shipping there." Moreover, the name honored the Reverend Nathaniel Ward of Ipswich, an uncompromising Puritan who sought refuge in the New World. He arrived in 1634 and spent his first winter in Ipswich. (His son, John Ward, was instrumental in naming the nearby town of Haverhill.)

Any similarity between the two Ipswiches today centers on access and position in regard to the sea. Otherwise, the characteristic New England town is similar to its mother town in name only. The East Anglian town of Ipswich in the county of Suffolk is today a busy port and industrial center, but with relics and reminders of a romantic past.

The Port of Ipswich, the biggest between the Humber and the Thames, handles millions of tons of cargo each year. There is the Wet Dock which can accommodate ships up to 275 feet long and Cliff Quay which can accommodate ships at any state of the tide. The busy port is continuously developing, and new facilities and changes are constantly taking place.

Old merchants' houses survive near the docks. The Neptune Inn in Fore Street, now a private residence, was a dwelling even earlier than the date of 1639 which appears on its facade. Nearby is Isaac Lord's house with an inner courtyard leading to a warehouse on the quayside. Relics of the Eldred House, which once stood in this district, are now in Christchurch Mansion Museum. Thomas Eldred was an adventurer who sailed through the Strait of Magellan, one of fifty men to survive a rough voyage around the world.

Another merchant of fame was John Chaucer, a vintner who took his name from the *chausses* carried as deck cargo on wine ships. The name was immortalized by his son Geoffrey Chaucer, the great English poet who, although a Londoner and a courtier, understood the wineries of Ipswich and the quayside on the Orwell, to which he referred in the Prologue to *The Canterbury Tales* in his description of the pompous Merchant with a forked beard:

> He wolde the see were kept for anythyng
> Bitwixe Middelburgh and Orewelle.

From the wealthy merchants and trade of the Middle Ages came the numerous medieval churches which are sprinkled through the town. In modern times, however, the churches are rapidly becoming redundant.

One closed church, St. Mary at the Quay, with a fine hammerbeam roof, has had several important memorial brasses removed to Christchurch Mansion Museum, including the famous Pounder brass engraved in 1525 by a Flemish artist. The merchant Thomas Pounder, wearing rich fur-trimmed gown, is shown with his wife beside him, while two sons and six daughters kneel at their feet. The town's great merchant and benefactor Henry Tooley is buried here as well with his wife Alice, and a wall brass of 1551 depicts them kneeling in prayer with their children. Tooley dealt in the fishing industry, importing of wines, and in canvas and fabrics including a new type of colored cloth called a "medley."

Also disused is the Church of St. Clement, close to the wharves and cranes of the dock and hemmed in by commercial buildings. Long known as the Sailors' Church, it was built in the fourteenth century when the port was at a peak of prosperity and served sailors and others who worked on the docks. Inside is the tomb of Thomas Eldred and a stone monument to Chevalier Cobbold. The Cobbold family is an important name in Ipswich, known as brewers and appreciated by imbibers. The church, with an extremely fine example of clerestory in the fifteenth-century nave, suggests medieval beauty in a seaside corner of Ipswich.

But the heart of Ipswich has largely lost its seafaring atmosphere. The busy center is around the Buttermarket, the street which once held stalls for the sale of butter and dairy products as well as poultry. The area also once held active churches. St.Stephens, with a lovely west tower overlooking the Buttermarket, and the nearby St. Lawrence are now disused. The spires and towers of old churches may now be overshadowed by new building blocks, but the pattern of the narrow, congested streets contains interesting old buildings everywhere.

The Ancient House

Ipswich is proud of its ancient house in the Buttermarket called the Ancient House. It is a gorgeous example of an ornate plaster decoration known as "pargeting." The sculptured symbols below four oriel windows represent the continents—America, Asia, Africa, and Europe. Australia was unknown when the house was built in 1567. For two hundred years it was the home of the Sparrowe family, ardent Royalists, who are said to have sheltered King Charles II in a secret room. Indeed, the interior remains fit for a king. In the rooms are such features as oak-paneled walls, a carved plaster ceiling, and a Tudor fireplace. The building now functions as the very fine Ancient House Bookshop.

Just as the Buttermarket is not restricted to the sale of butter, so Cornhill is no longer the place where corn is sold. Rather, public buildings are located in the wide space in the center of town known as Cornhill. In the square, the Corn Exchange adjoins the Italian-style Town Hall, built in 1868 with domed tower and clock with four faces. Next to it is the unusually ornate Post Office.

Also in the town center, on a busy corner, is the White Horse Inn, an old coaching inn of Dickensian fame where Mr. Pickwick lost his way in the corridor and had an interesting encounter with a "middle-aged lady, in yellow curl-papers." The white horse can still be seen over the doorway, unkindly described by Dickens as "some rampacious animal with flowing mane and tail, distantly resembling an insane cart-horse." Unlike Dickens, Americans thought enough of this coaching inn to build a full-scale replica of it at their World's Fair in Chicago.

St. Mary-Le-Tower is so named because it stood near a tower of the old town walls, along the line of the present road, Tower Ramparts. This principal parish church was largely rebuilt in the 1860s, and its present appearance reflects Victorian taste. Still, there is a fifteenth-century font, an oak pulpit of about 1700, and medieval misericords in the chancel. Parish record books of the seventeenth and eighteenth centuries contain minutes that suggest that meetings may have been too long, for they "ordered that this Meeting be adjourned to the White Horse Tavern."

St. Margaret's is more spectacular. Set against the background of Christchurch Park, it has a fifteenth-century double-hammerbeam roof, a beautiful clerestory, and attractive decorations all over. A font is carved with eight angels, one holding a scroll with the inscription "Sal et Saliva" referring to the use of salt and spittle in an ancient baptism ceremony.

Just north of St. Margaret's Church, in the spacious Christchurch Park, originally outside the old town walls, is Christchurch Mansion. The fine Tudor structure was built in 1548 by Edmund Withipoll, a London merchant. Among the features of the typical E-plan red brick building are rooms with Tudor paneling, hall with minstrel gallery, and kitchen with great fireplace equipped for work. Queen Elizabeth (who slept in more places than George Washington) is supposed to have slept in this house when she visited Ipswich in 1561 and again in 1565. Now filled with mementos of Ipswich history and maintained as a museum and art gallery, its collection of master paintings features works of those local artists, Constable and Gainsborough. Gainsborough started his career in Ipswich, where he lived for fifteen years following his marriage at the age of eighteen to an Ipswich girl.

Also in this period, an unknown actor, greatly encouraged by enthusiastic Ipswich audiences, went in 1741 to London. There, under the name of David Garrick, he made an immediate and sensational success.

Cardinal Wolsey, who may be the most distinguished person associated with Ipswich, was born here in 1471. He had already founded Christ Church College at Oxford when he began the building in Ipswich in 1528 of what was planned as a major school, intended to outdo Eton or for that matter anything at Oxford or Cambridge. But the grandiose plans for his native town were aborted and the building never completed. When Wolsey fell from power, his college fell too. Completely destroyed, only the simple red brick Tudor gateway with the arms of Henry VIII above the arch survives just alongside St. Peter's. Thus is the double fall described to Queen Katherine in Shakespeare's *King Henry the Eighth:*

Tudor Gateway of Wolsey's College

Ever witness for him
Those twins of learning that he rais'd in you,
Ipswich, and Oxford, one of which fell with him!
Unwilling to outlive the good that did it.

With so many relics of history, with the scale and quality of its many historic and ancient buildings, with a situation in unspoiled Suffolk countryside about sixty-eight miles northeast of London, it is no wonder that the words are still applicable with which Daniel Defoe described Ipswich over 250 years ago as "one of the most agreeable places in England."

King's Lynn

It is unlikely that a traveler would pass through isolated Norfolk by mere chance. Norfolk is an out-of-the-way county, characterized as flat (some would say "dull"), located in the area known as East Anglia, some one hundred miles north of London. So the visitor would normally have a reason, perhaps a business reason, for going to the port of King's Lynn. But no excuse is needed for enjoying the delights of a town which exudes charm and character and takes the visitor back to Olde England.

King's Lynn, known to the locals as just plain Lynn, has an ancient and engaging character that can be explained on the basis of geography. Located on the River Great Ouse which flows into the Wash (an inlet of the North Sea), it has always been an important port, and its wealth has always been rooted in trade. As a primary port of England in the Middle Ages, it vied with Boston (on the other side of the Wash) for the export of riches from the English interior, particularly wool. And it continues to be an important port.

Lynn, with medieval remnants and reminders of history and tradition, is one of the most romantic towns in all of England. The Guildhall of the Holy Trinity, with an exterior checkerboard pattern of flint and stone, dates from 1421. The interior displays a spacious Stone Hall with an impressive Perpendicular window and a beautiful Assembly Room. The Guildhall also houses many of the treasures of Lynn including the King John Sword and the King John Cup. King John visited Lynn on several occasions and is credited with having presented his own sword to the town in 1204. While the story is discredited that he presented the cup on the occasion of another visit when he lost his valuable baggage while crossing the Wash, the beautiful cup, gilded and enamelled with hunting and hawking scenes, is nevertheless an important treasure.

If the Trinity Guildhall can be recognized by its checkered front, St. George's Guildhall is distinguished for its checkered history, having served as a corn exchange, theatre, and warehouse. It was the theatre in Elizabethan times. Here, Shakespeare's company is known to have played, and it is likely that Shakespeare himself appeared on stage. Appropriately, it has been restored for use again as a theatre and is the center for the annual King's Lynn Festival of Music and the Arts. This marvellous festival, which has been taking place each summer since 1951, came about as an idea to help preserve the early fifteenth-century Guildhall. Towards the end of July, the finest orchestras, conductors, singers, instrumentalists, and actors perform in this "mini-Edinburgh" of the fens.

In Lynn, sea associations are strong and almost continuous. Along the quayside, in King Street, is a graceful Customs House, built in 1683 and designed by the local architect, Henry Bell. It is a square gem of two stories, surmounted by a lantern tower, with a statue of Charles II in a niche above the main entrance.

Customs House

120

Along the ancient waterfront are some of the finest medieval merchants' combined houses and warehouses in the country. A fourteenth-century wealthy merchant's house known as Hampton Court has been restored for use as flats. There is the long Hanseatic Warehouse of 1428, which extends back to the river. And the medieval Clifton House gives a good view of the town from its Elizabethan brick watch tower of five stories.

The towers of St. Margaret's Church overlook the old warehouses. Portions of this fine church, such as the southwest tower, date to the twelfth century. A mythical account explains the name of the church. St. Margaret is supposed to have slain an evil dragon. For that admirable deed, the pretty, lovable, pious Christian maiden won the unwanted love of Olybius, the Roman general. Naturally, she refused to marry him, and was, therefore, in the year 278, beheaded. Margaret, the tutelar Saint of Lynn, can be seen on the official seal, trampling the distorted body of the dragon while piercing his head with her cross.

Among the features to be viewed inside St. Margaret's are the Early English chancel with circular east window, elaborate reredos, fourteenth-century choir stalls with carved heads including those of Edward III and the Black Prince, and incomparable monumental brasses.

Norfolk has more memorial brasses than any other county, and the two largest brasses in England, both of Flemish workmanship, both nearly ten feet long, are located in St. Margaret's. The Peacock Brass depicts Robert Braunche, who died in 1364, with his two wives, enjoying a sumptuous banquet for King Edward III at which a peacock is being served. The other brass, memorializing Adam de Walsokne and his wife, is dated 1349, the year of the great plague, and is embellished with a rustic scene depicting corn being taken to a windmill.

The market day tradition is strong in King's Lynn where there are two markets, one for Tuesday and the other for Saturday. The Tuesday Market, founded in the twelfth century, is held in a large open area of three acres that takes on an especially cheerful and colorful appearance once a week but retains its festive charm always. The dominant Duke's Head Hotel, also designed by Henry Bell, is one of the pleasing buildings in the Tuesday

Market Place. One street, in the northeast corner of the large square leads to St. Nicholas Chapel with more medieval memories.

Even the market scene suggests certain sea associations; the fresh fruits and vegetables displayed here serve as a further reminder of the produce from surrounding farmland which still keeps the harbor busy. In fact, the very farmland has been reclaimed from the sea. From earliest times, when the first settlers began to inhabit the fens, they not only depended on river or sea for their livelihood, but they contended with the severe winter floods which were quite usual. The fens have since been drained and reclaimed for agricultural use. One unusual crop in this rich food-producing area is lavender. The large acreage of attractive color and aroma draws many visitors and not a few bees.

Associations with the sea continue to be strong in the still substantial fishing industry. Between the Alexandra and the Bentinck docks is the waterway known as the Fisher Fleet. Here the tiny fishing boats land with their catches. Here artists, photographers, and visitors gather to take in the local color and perhaps a few local samples. Lynn is the place to try fresh cockles—on the docks, straight from the sea.

Although the Red Mount Chapel stands far from the coast in a park-like setting called the Walks, it too manages to be bound up with the sea. Indeed, it owes its very existence to the sea. The unique wayside oratory was built in 1485 for use by pilgrims. Anyone taking a medieval vacation, a pilgrimage, was likely to leave from Lynn, the port for the shrine at Walsingham. The small, exquisite chapel in the upper story of this red brick, octagonal building is built in the fashion of a cross and has a delicate, fan-vaulted ceiling, among the finest extant.

King's Lynn is a delight for wanderers with an eye to making new discoveries and finds (possibly even King John's lost jewels). In just a small area bordered by the waterfront, something of interest exists in every street and at every turning. The narrow streets and fine medieval buildings convey an old world character, particularly along King Street, which leads from the Tuesday Market and changes its name to Queen Street after taking a slight bend.

In contrast to the romantic past, is the realistic present of the High Street, a modern, pedestrianized shopping street where one can buy new things (instead of discovering them) and, over a pot of tea, contemplate the changes wrought by time. For a much nastier shock back into reality, there is the heavily-trafficked John Kennedy Road.

Further afield, the variations are endless. The royal estates at Sandringham are nearby. Or there is Walsingham, that famous pilgrimage spot since 1061. There is the tiny timeless village of Castle Rising, once a Norman fortress complete with castle (as its name implies), with a perfect Norman church. Along the North Norfolk coast are some of the sandiest, sunniest, and loneliest beaches in England. And some forty miles south of Lynn is Cambridge.

So persistent are associations with the sea, that the very name of the town hearkens back to it. *Lynn* derives from a Celtic word meaning "pool" or "lake" and is thus a modest description of a marshy area in the Norfolk fenland once covered by the sea and floods. Early variants of the spelling include *Len* or *Lenn*, and the Norman scribe recorded it as *Lena* and *Lun* in the Domesday Book. During the Middle Ages, the town was known as Bishop's Lynn. But a 1537 charter of Henry VIII changed the name to Lynn Regis or King's Lynn, as it is officially known. The locals drop the formal prefix and refer to their town in the familiar form, "Lynn."

And that was the name chosen for the Massachusetts counterpart to honor the Reverend Samuel Whiting, who left his Church of St. Margaret in Lynn and embarked for the American wilderness just sixteen years after the Pilgrim Fathers set sail. There he worked for forty-three years, until his death in 1679. The American Lynn has long outgrown, commercially, its older and more humble namesake, the beautiful and intriguing town of King's Lynn.

Lancaster

The county town of Lancaster, made a city as recently as the time of George VI's coronation in 1936, has an ancient history. Its name derives from the Roman *Castrum*, or camp, built in Hadrian's time to guard the River Lune. (The remains of a Roman bath house and the wall of a corner bastion tower of the fort reflect former Roman dominance.) Angles called it Lunecaster, and Normans made it their headquarters.

The Normans built the castle, which was later enlarged by John of Gaunt, Duke of Lancaster as well as the "time-honoured Lancaster" of Shakespeare's *Richard II*. Elizabeth I added fortifications to the castle, which became a parliamentary stronghold in the English Civil War. And from the eighteenth century, it was used as courts and a prison. The thirteenth-century fortress with a Norman keep remains the cynosure of the city and continues to dominate the gray city below from its position on Castle Hill.

Adjacent to the castle stands the Priory and Parish Church of St. Mary, mainly Perpendicular in style, but dating back in part to the Saxon period. Of particular note are the rich carvings on fourteenth-century oak canopied choir stalls, among the finest in Britain.

All around the Castle Green are delightful eighteenth-century houses of attorneys and court officials. Facing the castle is the Cottage Museum of 1739, recently restored and furnished as an artisan's cottage. Elsewhere, particularly in Church Street, are fine Georgian houses.

The Friends Meeting House dates from 1690, a time when the Quakers established a strong presence in Lancaster; but the oldest house in Lancaster, the Judges Lodging in Church Street, was built in the 1620s by Thomas Covell and later used by judges during the Lancaster assizes. The house now accommodates two museums: the Museum of Childhood with displays of dolls, toys, games, and a Victorian schoolroom; and the Gillow Museum with displays of the history, tools, and products of the eponymous Lancaster furniture designer and other local cabinet makers.

Further examples of Gillow furniture are exhibited in Leighton Hall, the country home of the Gillow furniture-making family. Located in Carnforth, the beautifully situated house also offers to the public fine views of Lakeland fells and flying displays with trained eagles and falcons.

Located near the mouth of the River Lune, Lancaster was once an important port with links to the West Indies slave and sugar trade. In the center of St. George's Quay is the mid-eighteenth-century Custom House designed by Richard Gillow; it serves now as the Maritime Museum and confirms the city's former power as a port. Here on the quay, such staple West Indies products as sugar, cotton, molasses, and rum were unloaded in Georgian times, as were shiploads of mahogany, the raw material for Gillow furniture. A reminder of the golden age of maritime prosperity, the quay is nowadays a lively place with an annual maritime festival that centers on the museum and quayside pubs.

Another of the city's seemingly ubiquitous museums is housed in the Old Town Hall of 1783 in Market Square. The City Museum concentrates on collections of archeology and the social and economic history of the area.

For a pleasant interlude, the Lancaster Canal, with its splendid aqueduct over the Lune, provides fascinating towpath walks alongside houses, pubs, warehouses, and surviving textile mills that once took advantage of cheap canal transportation for their supplies of coal. Cruises on narrow boats or hired punts offer the opportunity to experience the canal from the vantage point of a pleasure craft and to appreciate, in particular, the five-arched aqueduct. Completed in 1797, this magnificent engineering feat was designed by John Rennie to carry the canal across the River Lune.

In Williamson Park is another attraction, the domed Ashton Memorial, a huge Edwardian folly and landmark for miles around. Lord Ashton completed "the structure" (as it is known by the locals) on landscaped grounds given to the city by his father, James Williamson. Family wealth based on a huge textile and linoleum industry also built the city's new Town Hall in Dalton Square. But the most useful thing about "the structure" may be the stupendous view from its galleries towards the Welsh mountains and northwards to the Lake District and to a nearby village with an American connection.

In the North Lancashire village of Warton lived family members of the first American president. Robert Washington built the tower of St. Oswald's Church over five hundred years ago, and some eighteen members of the family were baptized in the twelfth-century font. Thomas Washington, vicar from 1799 to 1823, was the last of the family to live in the village which recognizes the connection by flying the American flag from the church tower each year on the Fourth of July.

Situated only twenty-five miles from the Lake District (and 238 miles northwest of London), the city stands on a tidal river overlooked by the defensive castle with its massive gateway built for the Duke of Lancaster and son of John of Gaunt, Henry IV, who usurped the throne in 1399. As if in retribution, the name attached to so great a power was usurped by persons and places over the centuries and eventually adopted by cities across the ocean.

In Massachusetts, Lancaster was established as a town in 1653 on petition of the inhabitants of Nashaway. Lancasters in other states followed. In New Hampshire, for example, Governor Benning granted a charter in 1763 incorporating the town of Lancaster. Surely, names intensify the connection—wherever Lancasters occur.

Laxton

The scenery of Nottinghamshire—Sherwood Forest and Robin Hood legends notwithstanding—is dotted with collieries and factories. The midst of the Midlands is modern, smoky, and industrial, sustaining any impression of an unattractive coal mining industry that may have been fostered in the novels of D.H. Lawrence. What a jolt it is, therefore, to find an ancient village, not only wholly agricultural, but with a system of farming that is over a thousand years old.

The village of Laxton, known in earlier times also as Lexington, is in the center of Nottinghamshire on the eastern edge of Sherwood Forest. Narrow roads connect it to adjacent farming communities, but two major highways on either side bypass it and leave this quiet segment of the country isolated. Thus, the casual tourist is not likely to come upon Laxton and its intriguing past nor to realize that this ancient village is the ancestor of Lexington, Massachusetts.

What happens to names through the ages? Framlingham loses an *l* and Billericay loses a *y*. Maldon undergoes a spelling change. In England, Nottingham loses its initial *S* from the original Anglo-Saxon town of Snotingham. But Lexington loses a syllable and becomes Laxton. Largely disused by the early seventeenth century, the name "Lexington" is Saxon. The first part refers to the original head of the community, and the *-ing* root refers to "the descendants or people of." It is the tun, or farm, of the people of Leaxa.

An entry in the Domesday survey of 1086 refers to "Laxintune." Other early documents refer to such variants as Lexintun and Lexinton, Laxington and Lessington.

At the time of the Norman Conquest, Laxton was held by Tochi, son of Outi. But William the Conqueror granted it to Geoffrey Alselin, a great Norman baron. Other lords of the manor who succeeded him include Robert de Caux, builder of the castle; Robert and John de Lexington, judges and administrators;

127

their brother Henry, bishop of Lincoln in 1253; Robert de Everingham; and so on, to the the present owner of the estate, the Minister of Agriculture.

Laxton is the only place left in England where the open field system of agriculture is still practiced, using land tenure methods unchanged since Saxon times. Each farmer changes his land year by year. Three great fields of about three hundred acres each—West Field, Mill Field, and South Field—are further divided into strips of about three and a quarter acres to be distributed among the tenants. Yearly, in rotation, wheat is planted in one field and crops in another, while the third lies fallow.

Dovecote Inn

The system persisted as would have been natural under primitive conditions. A long narrow strip was most convenient for a plough drawn by a clumsy team of oxen—long to avoid having to make awkward turns more often than necessary, and narrow to limit the area to that which could be ploughed in a day. Strips were scattered to insure more equal distribution of good land among the farmers.

The manorial court which administers the system still meets annually. It appoints a jury to inspect fields and to insure that customs of the manor are observed. After their inspection, after they have driven in any new boundary stakes and made necessary repairs, the jury members repair to the Dovecote Inn for a repast. Refreshments are paid for in part by proceeds from fines that have been collected by the court.

After the harvesting of crops, the fields are open for common grazing, and stretches of meadow at the ends of the furrows are auctioned off for the right to cut hay. The bidding is open to those who, in the words of the ancient rules, "have smoke up a chimney in Laxton"—that is, to residents. Again, the ritual ends with refreshments at that aviary of activity, the Dovecote Inn.

Why this system of agriculture survived at Laxton is uncertain. That it will continue to survive is more certain, for the Minister of Agriculture is now lord of the manor.

A minister of fortifications might have saved the castle. Nottinghamshire castles are in general disappointing, for most have only the green mound remaining. Although this is true of the Laxton castle, it nevertheless survives as an unexcelled example of the motte-and-bailey type of Norman castle. A motte is the steep mound which forms the main feature of eleventh- and twelfth-century castles, and a bailey is the open space or court of a fortified castle. The mount at Laxton, with a circumference of 816 feet and a slope of 71 feet, was surrounded by an inner bailey of nearly eight acres. A short grassy walk from opposite the church leads to the remains of the ground plan.

Only the imagination can conjure up the splendid castle built soon after the Norman Conquest by the de Caux family, with a view encompassing Lincoln Cathedral some twenty-four miles away. Any disappointment that is felt may be tempered by two bonuses. First, the castle mound offers a close-up view of the fields. Second, it offers a distant view of the village and the dominating church.

The large, thirteenth-century Church of St. Michael went through a period of neglect and abuse in the eighteenth century when it was actually strewn with rubbish. Then those well-meaning Victorians "restored" it. The early English tower was dismantled together with the last bay of the nave. A new tower was erected and the nave shortened by one bay, thereby distorting the proportions of the building. Nevertheless, it is noteworthy for such splendid features as the magnificent nave clerestory, the tall and slender circular piers of the nave arcade, a fourteenth-century font, and the north aisle screen of 1532. Curious gargoyles outside include grotesque animals, a cross-legged satyr, and a crouching man.

Gargoyles on Church of St. Michael

Inside are the tombs of the Everingham and Lexington families, lords of Laxton from the thirteenth to fifteenth centuries. Of particular interest is an attractive monument of Sir Adam de Everingham of 1335 and an oak effigy of his second wife, the only surviving wooden medieval effigy in the country.

Even more interesting (at least to Americans) is an exhibition in the south aisle by local school children depicting village history. In a case containing an assortment of models, pictures,

130

and memorabilia, is a notice attached to the Declaration of Independence announcing that Laxton is also called Lexington. "The Lexington in America Massachusetts," it continues, "is called after this Laxton."

Evidence that Lexington in Massachusetts was named for Laxton is circumstantial. Originally known as "Cambridge Farms," the New England town was incorporated under the new name of Lexington in 1713. Since early settlers named most towns for the places in England from which they emigrated, it is reasonable to suppose that they applied this practice to Lexington as well.

It is extremely likely that settlers came from Laxton or the area. The nearby village of Scrooby is known, when it is remembered at all, for its Anglo-American ties. Postmaster William Brewster, born in 1566 in Scrooby, was a leader of the Separatist congregation which met at his home. Members of the group, including the famous diarist, William Bradford, came from the surrounding cluster of villages. This spirited band of Separatists, forced to flee to Holland, later became the group of Pilgrim Fathers who sailed to the New World in 1620. Naturally, they continued to feel the old world ties strongly.

Some historians claim that the newly-incorporated town was named for Lord Lexington. But since the title emanated from the manor of Lexington (Laxton), it is clear that the name of the town in New England derives one way or another from the small town in Old England.

Some cynics say that the English save everything and discard nothing. They have even preserved an ancient system of farming which managed to survive through centuries of vast social and economic changes. Well, the American Lexington has preserved in its name the reminder of the remainder of an ancient past, a tribute to a living museum. To complete the circle begun by those earlier pilgrims, how appropriate it would be for modern pilgrims to see the source and perhaps cap the visit with a stop at the Dovecote Inn (and the Laxton Visitor Center next to it). A refreshing anachronism in modern times is Saxon Laxton.

Leicester

Leicester was built on the Roman site of Ratae Coritanorum, meaning the ramparts or earthworks of the Coritani people. The Romans laid out a planned town in about the year 100, of which the most striking remaining remnant is a sizable piece of masonry known as Jewry Wall. Both its origins and its name remain a mystery, but it was probably the grand entrance to the palaestra, the exercise or games hall of the bath complex. The wall is located close to the oldest church in the city, St. Nicholas, which retains some work done by the Saxons who settled here after the Romans left.

Soon after the Norman Conquest, the invaders built a motte and bailey castle, of which the mound (30 feet high and 100 feet across) survives. The bailey contained the early twelfth-century Great Hall, its eastern wall rebuilt in brick in 1695, which remains the core of what is today known as the castle. John of Gaunt's Cellar still survives, but other buildings within the castle bailey have almost entirely disappeared.

The Norman castle at Leicester is reached from the north by a picturesque fifteenth-century black-and-white timbered gateway in Castle Street. From the south it is reached by the ruins of a fourteenth-century Turret Gatehouse, with a Norman half-timbered hall still in use as the Assize Court. The portcullis slots of this entrance to the castle bailey are still visible.

Outside the gateway is the Newarke, a fourteenth-century walled enclosure adjoining the castle, with a hospital for the elderly, houses for priests, and a church—all within the walls of the "New Work" which was swept away at the Reformation; but still remaining is the enormous fifteenth-century Magazine Gateway, so named because it stored munitions during the English Civil War. Close by is the Newarke Houses Museum, exhibiting social history of the past four hundred years. (Where else may visitors enrich themselves with a knowledge of the hosiery industry, which developed in the seventeenth century and continues to give the city fame?)

The Church of St. Mary Castro, which stood within the bailey of the castle, retains an abundance of Norman work with a Norman sedilia in the chancel and Norman doorways in the north and west walls of the aisle. The transept was extended in the thirteenth century and the south aisle was built with a tower at the west end. A fifteenth-century wall ran from the church to the Turret Gateway.

Leicester developed into a thriving medieval town with seven parish churches. Located in the center of an agricultural area, the market square has been in use since the thirteenth century. The small market town changed to a prosperous and major urban hub in Victorian times, symbolized by the Gothic clock tower of 1868 in the center of the city. The fifteenth-century guildhall, containing a timbered hall and a mayor's parlor, became too small to handle an increasing population, and the new Town Hall was built in 1874 in Queen Anne style.

The parish church of St. Martin's was elevated to cathedral status in 1926, and the county town of Leicestershire became a busy cathedral and university city.

Leicester, a large industrial center, gives the impression of space. The important and bustling medieval fulcrum has become in modern times a place for quiet contemplation in the midst of commercial turmoil. A final subject for meditation focuses on the element of romance attached to the city by those who assert that it is the legendary realm of King Lear, supposedly an ancient British king of the ninth century BC.

Less romantic is the notion that the city provided the name that graces the Massachusetts city of Leicester. It all began in 1686 when the Nipmuc Indian tribe sold a tract of land called Towtaid to nine men from Roxbury, Massachusetts. The deed was not recorded until 1713, when the township was approved and named Leicester "after the ancient City of that name in England."

Leominster

Once upon a time, hurricanes did happen in Herefordshire—at least in the sense of stormy history. Located as it is near the Welsh border, in a defensive position near the confluence of the Rivers Lugg and Arrow, Leominster found itself in the center of many upheavals.

Under the rule of the Saxon King Offa, a huge earthwork known as Offa's Dyke was built in 782 as a frontier line designed to give a measure of security to Leominster people who were in constant danger of raids from Wales. A good section of Offa's Dyke can be seen at Lyonshall, about ten miles west of Leominster.

The ninth century marks the beginning of severe struggles with the Danes; their armies ravaged the country, sometimes as a joint venture with the Welsh, probably destroying a Benedictine nunnery at Leominster in 980.

The first religious establishment had been founded in 660 with Ealfred, a Northumbrian missionary, as the first abbot. According to legend, the hungry Ealfred was eating bread when an equally hungry lion approached. He offered bread to the ferocious animal whose meek acceptance was taken as a certain sign that Ealfred would succeed in his work among the pagans. According to some, the legend of the lion is preserved in the name of the town, "Leonis Monasterium." More likely is the theory that "Leofminster," as it appears in Domesday Book, is a contraction of Leofric's Minster; Leofric, Earl of Mercia and husband of the better known Lady Godiva, had endowed a nunnery here, which replaced an earlier one destroyed by the Danes. But little is known of that nunnery, which was disbanded by 1046.

After the Norman Conquest of 1066, many castles were built along the Welsh borders, and a period of relative peace ensued during William's reign. But with his death, the Welsh resumed their incursions into England. In a particularly destructive raid of 1207, the town was severely plundered and burned and nearly totally annihilated. To fight off the freqent invasions, castles continued to be built, especially throughout the twelfth and thirteenth centuries.

The eleventh-century Priory Church stands as a link to the early history of Leominster. Rebuilt and extended during the prosperous thirteenth and fourteenth centuries, it is known today for its three naves. The central nave became the parish church in 1239, while the south aisle was added in the early fourteenth century. But the Priory was destroyed with the Dissolution of 1539 when the whole eastern end, including the central tower, was demolished; however, the naves have survived to become the parish church.

Seen today, the Priory Church of St. Peter and Paul may have just a portion of its former glory, but that glory is magnificent—featuring a Norman north nave with huge pillars sixteen feet in circumference, contrasting slender pillars in the south aisle, a thirteenth-century font, and a thirteenth-century wall painting of the wheel of life. Fourteenth-century floor tiles from the part destroyed in the Reformation are preserved in the west end. Also preserved is the finest surviving ducking stool in the country, used as punishment as late as 1813.

The town prospered in medieval times with the making of finely textured woolen cloth known as Lemster Ore. Poets have praised it, and the seventeenth-century poet Michael Drayton extolled it in his verse:

> Where lives the man so dull on Britain's furthest shore
> To whom did never sound the name of Lemster Ore,
> That with the silkworm's thread for smallness doth
> compare.

But his contemporary, Robert Herrick, found a "bank of moss more soft than the finest Lemster Ore."

Leominster remained a flourishing wool town until the Industrial Revolution, when the wool industry moved to the north. Today, the principal economy of the area is agriculture, and Hereford cattle are exported all over the world from Leominster. It is a quiet little market town with a population just over seven thousand, friendly and filled with plenty of picturesque old houses. Indeed, a profusion of timber-framed buildings make Leominster a handsome black-and-white town.

Draper's Lane, too narrow for cars, is an almost unspoiled pedestrian street which opens into Corn Square with a number of black-and-white buildings and a blending of various architectural styles. The Friday Market fills the space in the square once a week, and a walk leading past Lloyds Bank goes to the Grange, a large grass area with the town's ancient earthworks running along the edge. A distinctive structure at the east end, Grange Court, is now used for municipal offices.

Grange Court

Designed in 1633 by John Abel, the king's carpenter, the ornate building was originally the centrally-located market for butter and eggs. But the open space below the half-timbered upper floor was filled in; and the building itself, deemed a traffic hindrance, was dismantled in the nineteenth century and re-erected in its present position. Among numerous other buildings of special interest is the Forbury Chapel of 1282 and the fourteenth-century Grafton House.

Several nearby National Trust houses are open to the public. The neo-classical Berrington Hall has extensive grounds laid out by that supreme landscape gardener of the eighteenth century, Capability Brown. Five miles northwest of Leominster is Croft Castle, a Welsh border castle mentioned in Domesday Book; it is a reminder that this was once a land of turbulence in a time when many castles were built along the border for protection against raids.

If anyone emigrated from the finally peaceful and lovely town of Leominster to settle and name that other Leominster in America, the definitive facts cannot be ascertained. Nevertheless, there is more than a tacit relationship between the only two Leominsters known. An oil painting in the church suggestive of the surrounding countryside and entitled "Apple Blossoms for Leominster, England" is a gift from Leominster, Massachusetts, given in June 1976 in "the Bicentennial Year of the U.S.A. and the Thirteenth Centenary of the Diocese of Hereford."

Friendly Lemster (as the locals still pronounce it) is set in a beautiful valley among apple orchards and hopfields and lands in which plum-red Hereford cattle graze. Old timbered houses, market towns, and black-and-white villages add to the feeling of beauty and peace of the county. Indeed, hurricanes and harrowing history hardly ever happen.

Lincoln

The Cathedral of Lincoln dominates the city and the surrounding countryside. Situated high on a hill, the honey-colored stone cathedral is a spectacular sight from far or near. "A poem in architecture," it has been called. And Ruskin says that it is the best piece of architecture in the British Isles. The facade is richly decorated with a sculptured frieze over three elaborate Norman arches. A central tower of 271 feet is surpassed in height in England by the spires of Salisbury and Norwich. But in sheer beauty, many would agree, the cathedral is surpassed by none.

A previous cathedral, begun just half a dozen years after the Norman Conquest, was destroyed first by a fire of 1141 and then by an earthquake of 1185. In 1192, Bishop Hugh began work on the present and impressive three-towered cathedral, work which continued after his death in 1200 until its completion in about the year 1235. Events after his death brought about a further and major renovation. The Angel Choir at the east end was built between 1260 and 1280, replacing the apse end of the previous design, in order to accommodate crowds of pilgrims who came to visit the tomb of the canonized Bishop, St. Hugh.

Dubbed the Angel Choir because of the number of carved angelic figures, it made a suitable setting for the Saint's shrine. Thirty figures of angels in spandrels of the arches perform a variety of activities. One reads a scroll, another holds up sun and moon in his hands; many play instruments.

There too, in the choir, on one of the corbels, sits the famous Lincoln Imp. Cross-legged, wide-mouthed, elfin-eared, and a reminder of the less angelic aspect of life, he is a major attraction. With the magnificent Angel Choir, the cathedral was complete and completely magnificent.

Edward I and Queen Eleanor attended the dedication ceremony in 1280. The queen died just a few years later at Harby, ten miles from Lincoln, and her monument is in the Angel Choir. Lincoln had the first of a series of twelve memorial crosses

Lincoln Imp

erected by a heartbroken Edward I to mark each resting place on the funeral journey from Harby Church to Westminster, where she was buried. Also in the Angel Choir is the fascinating tomb of Robert Fleming, founder of Lincoln College, Oxford. Underneath his effigy on the upper tier is a skeleton reproduction, a reminder of the ephemeral nature of life.

Another tomb reveals the Little St. Hugh episode of 1255. When the boy's dead body was discovered in a well, the Jews of Lincoln were blamed for his ritual murder. They were persecuted and many were executed. The boy was canonized and immortalized in legend, as recalled by Chaucer in "The Prioress's Tale." The shrine in the south aisle of the cathedral is now accompanied with a disclaimer and apology for a totally discredited and shameful bit of Lincoln's past. It reads:

> Trumped up stories of 'Ritual Murders' of Christian boys by Jewish communities were common throughout Europe during the Middle Ages and even much later. These fictions cost many innocent Jews their lives. Lincoln had its own legend, and the alleged victim was buried in the Cathedral. . .

> Such stories do not redound to the credit of Christiandom, and so we pray:
>   Remember not Lord our offenses, nor the offenses
>   of our forefathers.

In the spacious cathedral interior, the nave is highlighted by dark Purbeck marble columns and by a twelfth-century font of black Tournai marble, embellished with figures of grotesque animals. The transept is adorned with stained-glass windows of the thirteenth and fourteenth centuries. A kaleidoscope of colors

Lincoln Cathedral

in the round window at the southern end is called the Bishop's Eye. An attractive medallion window known as the Dean's Eye is in the other end of the transept.

The Cathedral Treasury contains one of the four original surviving copies of the Magna Carta. A thirteenth-century cloister on the north side leads to the library designed by Sir Christopher Wren and to the polygonal Chapter House, a graceful, thirteenth-century building with flying buttresses and single central pillar. In the *yard* (a Saxon name for the precinct of a cathedral), is a statue of Alfred, Lord Tennyson, who was born in 1809 at nearby Somersby.

In the Seamen's Chapel of the cathedral, modern stained-glass windows commemorate Lincolnshire people who influenced the history of America. Specific reference is made to "Captain John Smith, founder of the Colony of Virginia and prime mover of the establishment of New England." Many of the Pilgrim Fathers who sailed in 1620 came from Lincolnshire villages, particularly around Gainsborough. When Lincoln, Massachusetts, became a town in 1754, it was given its name at the behest of the Honorable Chambers Russell, whose ancestors came from Lincolnshire. Early arrivals began a new history in a new land, after leaving behind an ancient and rich heritage.

Facing the cathedral is the east gate of Lincoln Castle. William the Conqueror understood the strategic advantage of this hilltop site overlooking town activities and the River Witham and ordered the erection of the castle in 1068. In a kind of modern urbanization program, 166 houses were razed to make room for the castle. The Observatory Tower, with an excellent view of the area from its top, was erected on a forty-foot-high mound. The Lucy Tower was built on a second mound. A special attraction is the lovely oriel window in the castle gateway, removed when John of Gaunt's Palace was dismantled.

The view from the tower confirms that not only the Normans, but all of the previous inhabitants of this hilltop site knew what they were doing when they chose the crest of the hill, a site which had been occupied in prehistoric times. Iron Age potsherds have been found as well as pre-Roman relics.

Oriel Window in Castle Gateway

First called Caer-Lindcoit by Britons and then Linn-dun by the Celts, it was Latinized by the Romans to Lindum, which, like London, meant "the hill fort by the pool." Apparently, the part of the city below the hill was a stagnant pool or mere. The British root "lindos" meaning "marsh" persisted through the evolution of the name.

The Celtic settlement was replaced when the Romans made it the legionary fortress of Lindum by about the year 47. The Romans may have chosen the site for its accessibility to the sea and for its strategic placement on a narrowing of the River Witham. They had the advantage of a good overview of the marshes, which might be crossed by attackers from the north.

By the end of the first century, it was made a *colonia* for the use of retired veterans of the legions and called Lindus Colonia. A carefully laid-out plan reveals the sewage system, aqueduct, buildings, streets, and colonnades of the important walled town. Fragments of the wall remain. The Newport Arch, the north gate of Lindus Colonia, is unique in Britain. Traffic leaving the city via the main road to the north still goes through this Roman gateway.

Medieval Lincoln was equally important. The king appointed it a staple town, making it a place for the public sale of wool where merchants had to buy and sell. Because wool shipped out of Lincolnshire had to pass through Lincoln, the town became very wealthy.

That wealth is still in evidence in the twelfth-century houses exhibited on a very steep hill, appropriately named Steep Hill. Near the top of the street, in the cathedral and castle area, is a sixteenth-century, timber-framed merchant's house which now serves as the Tourist Information Center. Visitors may arm themselves with brochures or maps before meandering down the hill, viewing the ancient houses, to the town below.

The Jew's House, a stone building of the late twelfth century with the original Norman doorway and rooms on the upper floor, is located next door to the Jew's Court, believed to have been the Jewish Synogogue. The Norman House, sometimes mistakenly known as Aaron's House, is another fine example of early domestic architecture. A Jewess who later lived here was accused of debasing coin and hanged in 1290, the year when Jews were expelled from England. Interspersed with the many houses on the hill, or located in them, is an endless array of antique and gift shops, as well as tea shops and restaurants. With the Usher Gallery serving as a worthwhile detour, the visitor may never make it at all to the city below.

Below, in the main shopping area, more ancient sights await. The Stonebow, spanning the High Street, is a sixteenth-century gateway with the Guildhall situated above. High Bridge on the High Street, is a medieval bridge lined with shops and with steps leading down to "Glory Hole" where original Norman stonework of the bridge can be seen.

Newport Arch

Along the High Street are three medieval churches—St. Benedict's with its eleventh-century tower, St. Mary-le-Wigford with a part-Norman and part-Saxon tower, and St. Peter at Gowts, located close to the Gowts (the Saxon word for channels or water-courses, as in *gut*).

By the end of the fourteenth century, Lincoln was in decay. The wool trade was transferred thirty miles away to Boston, which had the advantage of direct access to the open sea. The wool rush was over, and the thriving, active city of Lincoln became a dull place with a history marked by recessions and disasters. Plague in the late sixteenth and early seventeenth centuries diminished the population. Devastation occurred in the Civil War. In 1724, Defoe described Lincoln as "dead, decayed and dirty."

But in the nineteenth century, new industry, a productive countryside, and the coming of the railways brought about an enormous expansion. Lincoln today has in its lower part its share of factories, chimneys, smoke, traffic, and rows of dull dwellings—the accoutrements of industrial society. Nevertheless, the ancient city offers an endless wealth of wondrous sights.

Even when the visitor to Lincoln has started his tour, logically enough, with the cathedral, what better place to end than with the cathedral? Before leaving, the tourist would do well to turn back and take in the view, while considering the poetic tribute paid by J.B. Priestley:

> Few things in this island are so breathlessly impressive as Lincoln Cathedral, nobly crowning its hill, seen from below. It offers one of the Pisgah sights of England. There, it seems, gleaming in the sun, are the very ramparts of Heaven.

It is easy to despair of mastering all of the history and beauty of Lincoln; but that kind of despair can only be a source of joy.

Ludlow

The relationship of the Massachusetts town of Ludlow to its counterpart in old England is based on conjecture. In New England, the region formerly named Mineachogue, Outward Commons, Cow Pasture, and Stony Hill was incorporated in 1774 as the new district of Ludlow. Why the name of Ludlow was chosen is not recorded in the annals of history, but theories abound. Perhaps a better question might be, "Why not?" Surely, the ancient English town with a splendid castle had a name worthy of adoption.

Ludlow is a perfect center for exploring the county with the magical name of Shropshire. However, with its castle and church, with its black-and-white half-timbered houses and delightful seventeenth-century inn, with its alleyways and antique shops, with the opportunity for walks by the River Teme, and with an altogether rich history, the visitor may not be highly motivated to leave the town for exploration of rich pastures all around.

Willa Cather wrote that Shropshire "is surely the country for the making of poets if ever one was." It made A.E. Housman a poet, and Cather traveled to Ludlow in 1902 for the express purpose of seeking out the author who wrote of the land of lost content in *A Shropshire Lad:*

> When smoke stood up from Ludlow,
>   And mist blew off from Teme,
> And blithe afield to ploughing
>   Against the morning beam
>   I strode beside my team. . . .

And:

> Oh, I have been to Ludlow fair
> And left my necktie God knows where
> And carried half way home, or near,
> Pints and quarts of Ludlow beer.

While Cather did not encounter the poet in Shropshire, she did became enthralled with the town of Ludlow and with The Feathers, the major inn at which she stayed.

The Feathers dates from the Elizabethan period. A private residence in 1603, it remains a structural delight today. The three-story, three-gable building leans out over the street known as the Bull Ring, a reference to the medieval cattle market which once occupied the site. Its white facade is covered with timber framing and a fantasy of richly-carved designs of scrolls, heads, animals, or geometric patterns. The architectural critic Nicolaus Pevsner calls it "the climax of urban black-and-white houses in the county of Shropshire." Its equally picturesque name was conferred when it was converted to an inn in 1669 and named in honor of the Prince of Wales whose traditional badge is three ostrich feathers.

Ludlow's past importance is reflected in the castle. Built as a fortress in the eleventh century, the castle was continuously enlarged until it was abandoned in the eighteenth. The impressive ruin, with two towers on its ramparts, rises on the top of a cliff. Beyond the outer bailey and the stone bridge over the former moat, in the inner bailey, a circular Norman chapel stands by itself on the greenery, open and roofless.

Empty rooms of the Great Hall call out for the imagination to supply scenes of an ancient past when the castle was peopled with a succession of historical figures—Edward IV and his two princes, who were later taken to the Tower of London and murdered; Prince Arthur who came here with his bride, Catherine of Aragon, only to die in Ludlow a few months later; his younger brother who succeeded him as Henry VIII; his ill-fated daughter who lived in the castle as a girl and in later years earned herself the title "Bloody" Mary; Sir Henry Sidney who, as governor of the castle, had it remodelled during his tenure; his son Sir Philip Sidney, the Elizabethan poet-statesman who spent his youth at Ludlow Castle; and John Milton, the great poet whose masque was performed in the Great Hall.

Rising high above Ludlow and dominating the town is the huge Gothic tower of the Parish Church of St. Laurence. With the size and grandeur of a cathedral, the fourteenth-century church is notable for its misericords with exceptional carvings in high relief on the underside of these choir stalls and for its stained-glass windows. The large east window illustrates the life of St. Lawrence, Ludlow's patron saint.

With a population of about eight thousand, the small but lively market town encourages aimless wandering after visiting its principle attractions, the castle and the church. All around the central Castle Square, which presents a particularly bustling scene on market days, are gems of black-and-white buildings which attract the curious visitor, as do alleys and narrow streets with evocative names—Mill Street, Fish Street, Pepper Lane, Harp Lane, The Narrows, Quality Square.

Houses near Castle Square

Broad Street is lined with rows of fourteenth- and fifteenth-century half-timbered houses as well as elegant Georgian buildings. At the top end is the Butter Cross, a classical stone building of the eighteenth century, which leads to a museum of

148

local geology and history. At the lower end is Broad Gate, the only surviving one of the seven gates of the medieval walled town. Buildings along the street reveal such fine details as fan lights over doors or lead waterpipes engraved with dates or faces. Lower Broad Street leads to the fifteenth-century Ludford Bridge over the River Teme. From the bridge are fine views of the countryside all around as well as of the vast church tower high above the town.

The market town of Ludlow, 162 miles from the capital city, remains a haven of escape from the crowded London scene in one of the most beautiful towns in England.

Maldon

The road from London imparts a feeling of open space as it wends its way eastwards through the Essex countryside. The attractive agricultural landscape is occasionally broken by charming villages or country pubs, and an occasional farm house will announce the sale of free range eggs or King Edward potatoes. Then, nine miles past Chelmsford, after a total distance of some forty miles—Maldon.

Maldon has evolved from a Saxon past—both the town and the name. It was Maeldune, the hill with a cross. Located in a picturesque part of Essex (a name derived from the East Saxons), the site was selected for its defensive position on a steep hill above the broad estuary, at the mouth of the Blackwater.

Saxons and Danes vied for possession of the area. Edward the Elder successfully defended it against the Danes, defeating them in 920. But in the summer of 991, the Saxons were decisively defeated by Viking invaders in a battle in which the valiant Saxon leader Byrhtnoth was slain. The events were chronicled and immortalized in a contemporary Anglo-Saxon poem, "The Battle of Maldon," apparently written by a Saxon survivor. Thus the journalist sang of the death of Byrhtnoth in his Old English epic:

> . . .the grey-haired leader bade
> His men keep heart and onward press, good comrades
> undismayed.
> No longer could he stand upright, his eyes to heaven he
> bent.

Buildings of the Saxon period have survived. One of the oldest churches in England stands beside the sea where it was built in the seventh century. The chapel, fourteen miles east of Maldon in Bradwell-on-Sea, was built in 653 by St. Cedd, Bishop of the East Saxons. Actually, Bradwell has a curious mixture of old and new. The chapel stands by the remnants of a two-thousand-year-old Roman fort and a twentieth-century nuclear power station.

In Spital Road, about half a mile from Maldon's town center, is the ruin of St. Giles Leper Hospital, probably founded by Henry II in the twelfth century and turned over in 1481 to Beeleigh Abbey, which was largely destroyed at the Dissolution; the Chapter House is a remaining fragment that survives as an extremely attractive private house.

Moot Hall

In the town center, near the top of the steep little hill on which Maldon stands, is the Moot Hall, built in 1440 by Sir Robert D'Arcy. The Borough Council still meets in this pleasing brick structure with overhanging clock. A staircase leads from the Council Chamber to the roof, which affords a fine view of the Blackwater flowing and winding its way out to sea. Georgian buildings on High Street and Market Hill are reminders of former wealthy maritime trade.

St. Mary's is a further reminder of thriving port days. Located downhill by the waterside, near the Hythe (Saxon for wharf or loading place), it was built on a Saxon foundation and rebuilt around 1130. The Norman structure remains the basis of the present church. The lantern tower, with beacon to guide mariners, was rebuilt in 1636 and again in 1740.

Two ancient church towers dominate the Maldon skyline today. The fifteenth-century tower of St. Peter's still stands—just the tower. The church itself fell in 1665. The Plume Library was erected up against the tower by Thomas Plume who also left his valuable collection of mainly theological and scientific books when he died in 1704. Among the five hundred volumes he donated are Dr. William Harvey's *Circulatione Sanguinis* of 1649 and a first edition of Milton's *Paradise Lost.*

A remarkable thirteenth-century triangular tower, unique in England, is the oldest part of All Saints' Church. Modern statues, in niches on the buttresses between the windows, represent such notables as St. Cedd (Bishop of the East Saxons), Byrhtnoth (slain Saxon hero of the famous battle), Sir Robert D'Arcy (builder of the Moot Hall), and Dr. Robert Plume (Archdeacon of Rochester and native of Maldon, whose library bears his name). The Vicarage behind the church is a charming fifteenth-century timber-framed house with ancient doorways and medieval passages.

Inside All Saints' Church are monuments of people who could have made the Guinness Book of Records. Edward Bright, billed as the "biggest man in England," weighed over six hundred pounds when he died in 1750. A monument of 1602 is to the memory of Thomas Cammock, a local tradesman, who had two wives and twenty-two children. After he eloped with his second wife, the couple escaped her father's hot pursuit by swimming their horses across the river to be married here in All Saints'.

But the feature for Americans is the Washington window, a gift presented in 1928 by the citizens of Malden, Massachusetts, to the memory of the great-great grandfather of George Washington. Lawrence Washington, loyalist clergyman dismissed as rector of nearby Purleigh, was buried in this churchyard in 1652. Glowing stained-glass colors in the window depict such scenes as the landing of Columbus in America, the arrival of the Pilgrims, and the first President of the United States taking the oath of office. An appropriate gift, it is also an affectionate acknowledgment of the founding of Malden by Joseph Hills of Maldon.

Throughout history, Maldon has owed its success to the sea. A main port in medieval times, old Maldon is today a principal yachting center, while the new Malden owes its existence to those who crossed the sea from the oldest borough in Essex.

Needham Market

The name of a town can be said to be an early written record of
the place itself and a source of information. The Saxons gave
the name of "Needham" to a home in need, that is, to a place of
refuge. Seen today, it is a place in need of refurbishing. It is a
scraggly town which straddles the main highway from
Stowmarket to Ipswich. And if it once held an important market,
there are no vestiges in the uninspired shops on the main street.

"Needham is but a poor Town," wrote the Reverend Thomas
Cox in the 1720s in his revision of Camden's *Magna Britannia*.
And Thomas Fuller said in 1662: "They are said to be in the
Highway to Needham who haste to poverty." But when
wool-combing was the staple industry, this little Suffolk town
was well off, and the church is evidence of former prosperity.

Set alongside the main street, parallel with the road, the
fifteenth-century Church of St. John the Baptist is made of flint
and stone, the usual East Anglia materials. Its exterior is
disappointing. But its interior is distinguished by a grand
hammerbeam roof, generally accepted as one of the best in
England.

Timber brackets, or hammerbeams, projecting out at right
angles carry the arched braces which support the heavy weight
of the roof. The need for a horizontal beam is thus eliminated,
and a spectacular architectural effect is achieved. Here, the
arched braces are concealed by a carved wood covering with
angels, and angels also embellish the hammerbeam ends.

Called "the culminating achievement of the medieval
carpenter" by Cautley in his *Suffolk Churches*, the remarkable
roof is worth studying to figure out how it stays up, for it gives
the final impression of a whole building suspended.

Needham Market has been loved by many for more than its
church. Thomas James, born in Boston in Lincolnshire, was
expelled from Needham in about 1661 under the Act of
Uniformity, but apparently he and his family retained their

fondness for Needham and its environs. His son Thomas, minister of Easthampton, Long Island, died in 1696, having stipulated that he be buried with his head pointing eastwards that he might face his people eternally.

That Needham in Massachusetts was named for Needham in England is clear. The evidence according to a statement by the Reverend Stephen Palmer in 1811 is that the town incorporated in 1711 "was named Needham at the request of Governor Dudley, after Needham in England; and because that town is near to Dedham though in a different county."

But which Needham? Several exist in England. A map of East Anglia discloses that the nearest Needham to Dedham in Essex is Needham Market in Suffolk, about twenty miles away. And continuing in a nearly straight line a further twenty miles northeast is Needham in Norfolk.

But Needham Market, with its wealthy church and past history and closer location to Dedham, is undoubtedly the actual mother town. The map is dotted with names that have been adopted in America—Sudbury, Acton, Ipswich, Framlingham, Braintree—and just to confound the issue is a name that is thankfully overlooked—Needham Street in Suffolk.

Newark

An ancient city with a castle in which King John died in 1216, the English city of Newark is resplendent with venerable history and culture. Not the New Jersey Turnpike, but the Roman Fosse Way carries the traveler across the county. Not the Raritan, but the River Trent meanders and links it with the more familiar city of Nottingham. And at its edge lies, not a major airport, but a major element of ancient legend, Sherwood Forest. Only in its name is the English city of Newark, full of medieval character, reminiscent of its New Jersey counterpart.

Visitors to this part of England are most often in search of the city of Nottingham or Robin Hood's Sherwood Forest. Both are easily accessible from the less frequented Newark and worth a visit, but worthy Newark is a delightful town with a great deal of history and with architectural gems to delight visitors.

Its name is believed to derive from New Ark or New Work—that is, new building—possibly a new town built on the ruins of the old. Emigrants may have conferred the name of their original habitat on the Puritan settlement they created in the New World, but the definitive source is lost to posterity.

The river that flows through the town to give it its full name, Newark-on-Trent, once also gave its name to a certain city that became the the capital of New Jersey. Now the River Trent carries colorful traffic beneath the walls of Newark Castle.

On the banks of the river stands the impressive ruin of the Norman castle, built in about 1129 and destroyed during the English Civil War when the local people defended Charles I. The north and west walls of the vast greystone fortress survive virtually intact and offer good views from ruined windows.

In the river below, pleasure boats make their excursions in the summer months, and a riverside walk leads to Millgate Folk Museum, a former oil seed mill, which features as part of its exhibition a street of nineteenth-century shops, rooms of a

The Parish Church of St. Mary Magdalene

Victorian house, a replica pub with hand drawn beer pumps, and a World War II shelter.

Not far from the castle and the river, in the center of Newark, is the cobbled Market Place, a large square area befitting a busy market town in the midst of the fertile and agriculturally rich Trent valley. Rich also with ancient architecture and history, the square is dominated on one side by the eighteenth-century Town Hall, with its stone Palladian front, and by the tall spire of the medieval parish church of Saint Mary Magdalene soaring over the scene from another side. Buildings all around conspire to give the feeling of the former medieval town center.

At one corner, the timber-framed former White Hart Inn remains a splendid example of fifteenth-century architecture with a restored facade that shows off a gallery of twenty-four painted figures under canopies. At another corner, the Governor's House, of an even earlier century, functions now as a bakery and repays a stop for a cup of coffee with a chance to view upstairs rooms that still contain medieval wall paintings and a well-preserved medieval toilet.

In between are old inns with literary associations. Byron was a guest at the Clinton Arms in 1806 when he visited the publisher who printed his two volumes of youthful poetry in a bookshop across the square, now Porter's bacon shop. The actual wooden press is preserved in the Newark Museum.

Next to Byron's haunt is the Saracen's Head, where Sir Walter Scott stayed several times and which he appropriated as a resting place for Jeanie Deans in *Heart of Midlothian.*

The square is especially vibrant on market days when hundreds of stalls with colorful canvas rooftops purvey a wide assortment of goods ranging from clothing and bric-a-brac to pots and produce. The dominant Town Hall has recently been refurbished at a cost of four million pounds with a pleasing ground floor shopping complex that goes through to the adjoining Buttermarket. From certain shops on the right side, it is just possible to distinguish cells of the former prison block where inmates were detained, while upstairs guests were entertained in an Adam-style ballroom. The lavish room, decorated in strong shades of pink and blue, and enriched by gold accents and designs all around, has apsed ends, square columns with Corinthian capitals, and a great central brass chandelier.

Everything circles around the square. Bridge Street leaves it for Appletongate where the municipal museum, located in part of the sixteenth-century Magnus Grammar School, houses a collection of prehistoric, Roman, and Anglo-Saxon artifacts.

But the most prominent period for visitors to Newark is the late Norman period, when Robin Hood and his merry men reigned. It is probably not possible to come to this part of Nottinghamshire and avoid reminders of the popular folk hero.

Although only a portion remains of the original Sherwood Forest which once stretched northwards from Nottingham to Newark for over twenty miles, the woodland sanctuary is filled with Robin Hood associations. The largest part of the surviving forest is near Edwinstowe, a village some fourteen miles to the west.

In the Church of St. Mary at Edwinstowe, Robin supposedly married Maid Marian. Around the huge oak tree north of the village he met with his band of a hundred men. Nowadays, countless bands of tourists pour in, particularly on fine summer Sundays—children capped with green Robin Hood hats and armed with bows purchased in the gift shop—to marvel at the Major Oak, which has a waistline of nearly forty feet. But the sixth largest oak tree in Britain was named for Major Rooke, a local eighteenth-century historian, rather than for its enormous size.

The hugely-proportioned Friar Tuck lived a mile west of Blidworth, home of Will Scarlet, who is buried in the village churchyard. But Friar Tuck's cell is purportedly at Copmanhurst, a hamlet so deep in the Nottinghamshire woods that it eludes the visitor. Indeed, many elements of the legend which has for centuries captured the imagination of countless numbers continue to elude even the most dedicated sightseers who come in search of the haunts of a hero whose story belongs largely to the realms of the imagination. Actually, even the famous oak—since oak trees live only five hundred years—is merely a symbol, serving as proof that great oaks from little legends grow.

Nevertheless, for a serious introduction, it might be best to begin at the Sherwood Forest Visitor Centre near Edwinstowe, with its audio-visual programs, talks by rangers, dining room, and inevitable gift shop. The center makes a good embarkation point for woodland walks and nature trails.

Other interesting villages abound in the Newark area. At Elston lived Erasmus Darwin, grandfather of Charles, and Darwin family monuments embellish the thirteenth-century church. Laxton is unique for the best preserved motte-and-bailey castle in the country and for its medieval system of open field farming, as exhibited at the Laxton Visitor Centre next to the Dovecote Inn. At Coddington, church windows are designed by William Morris.

Another literary site beckons. Newstead Abbey, built by Henry II in 1170, became the Byron family home in 1540. Open to the public, it displays many treasures including the table at which Lord Byron wrote some of his early poetry.

It was from the small cathedral town of Southwell (pronounced Suth'all by the locals), eight miles to the west, that Byron came to Newark. He was there frequently, especially on vacations from Harrow and Cambridge between 1804 and 1807, when his mother lived at Burgage Manor.

The beautiful Southwell Minster, mainly Norman work of the twelfth century, is known for stone carvings of various types of foliage decorating thirteenth-century columns and doorway of the chapter house. With magnificent twin towers on the west front, and with such stunning elements as a carved Saxon lintel in the transept, a bit of Roman pavement on one side of the transept, and a Roman wall painting, it deserves to be better known among visitors in search of the great cathedrals of England.

Easily of equal importance to Brits is the beloved Bramley apple, which had its inception at Southwell when it was propagated by a local nurseryman whose descendants still run a garden center exhibiting, naturally enough, a history of the famous fruit. A pleasant pub named "The Bramley Apple" pays tribute with its name, and various restaurant or tearoom menus may cap a visit with a fortuitous offering of a Bramley apple dessert known as Southwell Galette.

But a further literary offering may be tasted in Eastwood, where D.H. Lawrence was born in 1885 in a house in Victoria Street that is now open to the public. The family moved to several other houses in Eastwood, and the lad grew up in surroundings in which, he declared, "Robin Hood and his merry men were not far away." The rural landscape, with its coalfields and collieries, gave Lawrence the background for the early novels based in Nottinghamshire.

The River Trent creates a valley which links the industrial city of Nottingham (known for its university, theatre, and lace) to Newark—and it links Newark to its counterpart in the United States. With its castle on the river, cobbled market place, ancient buildings, and countless connections (Lady Godiva owned the town in 1050, Gladstone made it his headquarters in 1832, George Eliot stayed in 1868. . . .) old Newark makes an exciting focus for transatlantic visitors.

Oxford

Oxford in Massachusetts has existed since 1683 when the settlement of Nipmuck was given a new name to honor one of the greatest centers of learning in the world and the place where many Pilgrim Fathers had been educated. It is undoubtedly the same reason that accounts for the existence of an Oxford in a number of other states including Maine, Connecticut, and Mississippi. But the new Oxford in Massachusetts was settled primarily by bands of French Huguenots seeking refuge from religious persecution. They gratefully retained the English name of their haven in the new world.

The first group of French settlers produced their own familiar names. A descendant of Pierre Beaudoin, James Bowdoin, became Governor of Massachusetts. Benjamin Faneuil was an ancestor of Peter Faneuil, that benefactor of Boston whose name was given to Faneuil Hall. Descendants of André Sigournais, the Sigourneys, are known throughout New England.

It is an interesting sidelight that the name of the neighboring town of New Roxbury in Massachusetts was changed to Woodstock to parallel English geography which has the town of Woodstock as a neighbor to Oxford.

There, with the names, similarity ends. The English city of Oxanforda dates to the eighth century with the founding of St. Frideswide's Nunnery. While there is disagreement as to the origins of Oxford University, it is known that a university, among the most important in Europe, was well established by the start of the thirteenth century. Oxford today is a factory city, mainly for automobile production, with sprawling suburbs. But the heart of the city contains the colleges in an area of enormous interest and beauty.

A short visit can be only superficial at best. Within a square mile area are listed 635 buildings of historical or architectural merit. Highlights and pleasures of Oxford can be best derived by the simple means of looking—walking and looking. Unlike

Radcliffe Camera

Cambridge with its open stretches of greenery and exposed buildings, Oxford compels the visitor to go through gates and get behind walls to experience the scenes that emanate from a university so ancient that its founding date is unknown. Perhaps the sight that a see-er should begin with is Radcliffe Square.

The scene is one of utter charm and harmony. The spire of St. Mary's rises from the High Street side of the Square. The dull yellow buildings of Brasenose and Exeter are on another side, and Hertford and All Souls are opposite. In the center, dominating the picturesque scene, is the dome of the Radcliffe Camera, part of one of the world's largest libraries, the Bodleian. The Bodleian—affectionately known as "the Bodley"—completes another side of the square. Undergraduates may be seen scurrying in and out of the library buildings, no books under their arms, for the Bodley is not a circulating library.

Inside, visitors may view an engaging exhibition on the ground level which includes such literary finds as letters written by Virginia Woolf, George Bernard Shaw, John Masefield, Florence Nightingale; an assortment of original manuscripts; Shakespeare's First Folio of 1623; memorabilia such as Shelley's guitar; and a variety of beautifully illuminated medieval manuscripts. Upstairs is the most ancient part, the fifteenth-century Duke of Humphrey's Library. Book shelves jut out into the room and form alcoves for readers' desks, an arrangement designed to take maximum advantage of natural lighting. Stained-glass windows and hushed whispers of attendants contribute to the air of solemnity that pervades.

Near Radcliffe Square is Sir Christopher Wren's Sheldonian Theatre, used for degree-giving ceremonies and other academic functions. A fine view from its cupola of the spires of Oxford and the streets below suggests that Oxford, subtitled a City of Spires, could accurately be retitled, a City of Aspiring Students.

Also near the Square, in Broad Street, is Blackwell's, the large Oxford book shop which caters to students. It is proud of its tradition which permits the prospective buyer to browse unmolested. Some fast readers have never had to buy any books!

Next door is Trinity College, notable for its fine baroque chapel, a quadrangle designed by Wren, and a lovely garden, perennially viewed by the bust of a famous alumnus, Cardinal Newman.

Discoveries of all kinds await beyond the gates of each of the more than forty colleges. The chapel of New College is

Tom Tower, Christ Church College

memorable for its impressive seven-foot statue of Lazarus by Jacob Epstein, stained-glass windows designed by Sir Joshua Reynolds, El Greco painting of St. James, important brasses, and quiet cloisters. There are also Alice-in-Wonderland associations and the cathedral of Christ Church College, the Shelley Memorial in University College, and the deer park of Magdalen.

Magdalen College Tower

Diversions, easy to come by in Oxford, range from a number of friendly pubs to punting on the river. Or visiting the Botanical Gardens. And even if Oxford were somehow divested of its colleges, the Ashmolean Museum—with its vast collection of antiquities, Eastern arts, tapestries, and paintings by such noted artists as Guardi, Tiepolo, Tintoretto, Uccello, Constable, and Van Dyke—would offer sufficient justification for the fifty-mile journey from London.

Add to the inexhaustible wonders of Oxford the innumerable attractions of the surrounding countryside—Blenheim Palace, the Cotswolds, ancient villages and churches, Stratford—and the more-than-casual visitor might need centuries to see it all and pay proper tribute to the prototype for all those American cities named Oxford.

Plymouth

The Pirates may have come from Penzance, but the Pilgrims did *not* come from Plymouth. Plymouth was the last port of call for the Pilgrim separatists only because of some bad luck. Although the *Mayflower* passengers were mainly from East Anglia, they had been settled for the past twelve years in Leyden in Holland. Now, on their way to the New World, the accompanying ship, the *Speedwell*, needed repairs; in Plymouth it was found to be unseaworthy and eventually left behind. Passengers dropped out or regrouped, and the *Mayflower* sailed on 6 September 1620, with a party of 102 aboard. The momentous occasion is commemorated in England by a plaque on the barbican and in America by the naming of their settlement after the English town which treated them so well.

Actually, there are some forty Plymouths in various parts of the world, a fact which suggests that Plymouth was used as a port of embarkation for many expeditions. Its coastal situation in the southwestern corner of Devon where it guards the English Channel gives Plymouth a position of great strategic importance. As its sea power developed through the centuries, it became a center for voyages of exploration and discovery and for naval maneuvers.

Nine hundred years ago it was a small village with the ancient name of Sudtone—South Town. The ancient harbor is still called Sutton Harbor. And the oldest part of the city is in that area, where a barbican or outer fortification of the castle of Plymouth once existed.

Plymouth's history is based on a tradition of seamanship and shipbuilding. In Norman times the Domesday village began to prosper as a fishing and trading port. Edward I was among the first to recognize its potential as a naval base. It was he who assembled a fleet of 325 ships in 1287 for the Bordeaux wars against Phillip of France. By the fourteenth century, the descriptive name of Plymouth came into general use for the town

Plymouth Harbor

situated near the mouth of the River Plym. It had grown from a
hamlet to the fourth most important place in the kingdom.

Plymouth thrived and grew rich during the years of the
Spanish threat. Several sixteenth-century merchants' houses
remain to attest to that earlier wealth. And the old character with

narrow, winding, cobbled streets remains as well. The oldest street in Plymouth is New Street, so named because it was new when it was laid out in Elizabethan times. The Elizabethan House in New Street has been restored and is open to visitors.

This was the age of merchant-adventurers, explorers, and pirates, with some Elizabethans being all at one time. The Hawkins family led in the competition among merchant seamen to secure the wealthy new trade for themselves. The notorious John Hawkins organized one of the first large-scale slave trades, selling African natives to the Spaniards in the West Indies. Drake joined the ventures of the Hawkins family, to which he was related, in privateering, plundering, and illicit trading—all aimed against the Spanish, and therefore quite acceptable.

Sir Francis Drake is probably Plymouth's most notable celebrity. His exploits enabled him to return to Plymouth laden with enough Spanish gold to ingratiate himself into royal circles. And in 1577, he undertook his famous three-year voyage around the world.

His statue enhances the wide grassy expanse overlooking Plymouth Sound known as the Hoe—a "hill" or "height." From this plateau high above the barbican he watched for the approach of the Spanish Armada in 1588, an invasion intended by King Phillip of Spain to remove Elizabeth from her throne and establish Spanish dominance of the New World—the greatest danger England ever had to face again until 1940.

The Hoe has a fine sea view and a fine bowling green, and there Drake played perhaps the most famous game of bowls in history as he waited for arrival of the vast Armada. When word came to him, Drake is supposed to have answered that there was "plenty of time to finish the game and beat the Spaniards after." There was and he did. The cool, if legendary, statement was not mere idle boast. Plymouth was Drake's home port, and he knew that the British fleet could not move out of the Plym estuary for some hours, until the tide had turned. He defeated the Armada and was immortalized by a grateful England.

Plymouth stood against Charles I in the Civil War of the 1640s. After Charles II came to the throne, he built, in 1666, the citadel. The fortification was designed not to defend England against foreign enemies but to intimidate the inhabitants he so completely distrusted. The northern side of the fortress has an abundance of gun ports and a commanding position of the town itself.

The Hoe

The citadel dominates all of Plymouth from its position on the eastern heights of Plymouth's great attraction, the Hoe. Nearby is the Marine Biological Association with laboratory and aquarium. On the west side of the lovely stretch of greenery is a cliff with panoramic views. The Eddystone lighthouse, fifteen miles out to sea, is visible. But Smeaton's lighthouse, on the eastern end of the Hoe, is visitable. It originally stood lighting the Channel from 1759 to 1882. But after the sea had undermined the rock on which it was standing, the lighthouse was dismantled and re-erected on the Hoe.

Plymouth, the largest city in the West Country and the home of the Royal Navy, was very severely bombed in World War II. Its center is almost entirely new and has been rebuilt with wide

171

streets and modern shopping centers. The Civic Center and Guildhall are just two new buildings in Armada Way. A remarkably beautiful job of restoration renewed the medieval Church of St. Andrew, with stained-glass windows in the east end by John Piper and a memorial window in the west to Lord Astor; he and his American-born wife helped enormously in the rebuilding of Plymouth.

Fortunately, many authentic old bits of Plymouth remain. The fifteenth-century Prysten House behind the church was probably the home of the monks of Plympton Priory who had control of the area before the Dissolution of the Monasteries. The Church of St. Budeaux on the outskirts of the city is the place where Sir Francis Drake married Mary Newman. There is the citadel in the city, of course. And the tiny streets around the harbor remain unchanged.

Plymouth has treated its visitors well. One notable arrival took place on 2 October 1501, when the fifteen-year-old daughter of Ferdinand and Isabella, Katherine of Aragon, landed at Plymouth to be married to Prince Arthur. The town cheered the princess who stayed for two weeks and never returned. Indeed, the ill-fated princess might well have wished she had never seen England at all, for the following November she was married in London to Prince Arthur and widowed within five months. Her troubles began when she became the wife of his younger brother Henry VIII and the first of his six queens.

Pocahontas, the beautiful Indian maiden who had saved the life of Captain John Smith, landed in Plymouth in June 1616. She was about twenty when she arrived with her husband, an English widower and tobacco planter, John Rolfe. She was lionized, taken up and entertained by fashionable circles until, alas, the English climate impaired her health. She died of consumption in 1617.

In 1762, the painter Sir Joshua Reynolds, who originally came from Plymouth, hosted Dr. Johnson on a tour of Plymouth lasting several weeks. Highly honored by Plymouth society, Dr. Johnson is reported to have gorged himself on tea and Devon cream and to have enjoyed the visit enormously.

In 1722, James Cook departed from Plymouth on his circumnavigation of the world. And in more recent times, in 1966, Sir Francis Chichester sailed from Plymouth on his single-handed journey around the world.

With all of these comings and goings, by sea and by land, Plymouth knows how to be a good host. Thousands of visitors come each year from overseas. The Pilgrim Fathers were "kindly entertained and courteously used by divers friends there dwelling," says the bronze tablet on the barbican. And Plymouth continues to extend kind entertainment and courtesy to modern pilgrims whether they arrive to play bowls on the Hoe or to pay homage to the past.

Rochester

Not too far from London—thirty-four miles southeast—in the county of Kent is the cathedral city of Rochester, known and exploited for its associations with Charles Dickens. But the city is also outstanding for its cathedral as well as for a towering Norman castle, perhaps the finest in southern England.

The castle, with the tallest Norman keep in England, stands in a magnificent position on a grass mound within the old city walls. Henry I built it to defend the Medway crossing at a strategic point where the river bends and broadens into its estuary as it winds its way out to sea. With four floors and a height of 120 feet, it offers from the top of its defensive tower a panoramic view of the city which takes in the cathedral. Mr. Pickwick visited the scene so enthusiastically praised by Mr. Jingle in the Dickens novel: "Ah! fine place . . . glorious pile—frowning walls—tottering arches—dark nooks—crumbling staircases—Old cathedral too—earthy smell—pilgrims' feet worn away the old steps—little Saxon doors. . . ."

The old cathedral dates from the Norman period, but the outline of the earlier Saxon church founded in 604 is visible. The present cathedral was built in the eleventh century by Bishop Gundulph, who also built the castle at Rochester, as well as the Tower of London. Completed in 1130, the cathedral is a place of great beauty and architectural variety, with an arched Norman crypt, magnificently proportioned Norman nave, and spectacular west front with round-arched Norman door with pillars and carved animals.

Near the junction of nave and crossing are the worn steps eroded by pilgrims climbing their way to the shrine of a baker from Perth, murdered in 1201 while on the start of a pilgrimage to the Holy Land. The monks may have established the shrine of the good baker (who gave every tenth loaf to the poor) to compete with that of St. Thomas in Canterbury. Miracles were reported, and St. William's tomb became another popular pilgrimage place, yielding enough money for the rebuilding of the entire eastern end.

Rochester Cathedral

Among the many memorials in the cathedral is, predictably, a tablet commemorating the great Victorian novelist whose associations are everywhere. After London, Rochester is the city which figures most frequently in the works of Dickens, sometimes appearing as Dullborough or Cloisterham. A walking tour of the city takes in "the Bull Inn in the High Street" (now the Royal Victoria and Bull), credited as "good house, nice beds" in *Pickwick Papers*, and appearing again in the High Street of Cloisterham in *The Mystery of Edwin Drood*.

Among other appealing sights in the restored High Street is the seventeenth-century red-brick Guildhall, now the main museum and once the place for local government and for market stalls which sheltered in its columned ground story.

Further along the main street are the Corn Exchange (of 1706) with a clock jutting out over the street, Chertseys Gate (a fine fifteenth-century gateway to the cathedral precinct), Blackboy Alley (a route leading to the cathedral's north transept, used by pilgrims visiting the shrine of William of Perth), the Six Poor Travellers' House (a sixteenth-century almshouse with six bedrooms and communal dining room which provided free accommodation for one night only), and the medieval city walls (which follow the line of Roman fortifications). Inevitably, the route defers to the city's obsession with the famous author. The Tudor brick Eastgate House contains the Charles Dickens Centre with illustrations, letters, and portraits displaying the novelist's development. In the garden stands the Swiss-style chalet taken from nearby Gad's Hill where the novelist lived and worked from 1856 to his death in 1870.

Just off the High Street is Restoration House, where Charles II stayed in 1660 on his way to London to be crowned. Two additional old houses worth seeing are in Boley Hill. Satis House ostensibly received its name when Queen Elizabeth, asked if she had been comfortable after spending the night there, replied "Satis." Next door is the timbered Old Hall where Henry VIII met Anne of Cleves on her arrival in England.

With Dickens associations employed to the fullest, it is easy to forget ancient associations when the small pre-Roman town by the crossing of the River Medway was Durobrivae—a name derived from the ancient British word *dour* (water) and *briva* (bridge or crossing point). Its appearance as Roibis in the Roman military tables is a short step towards Rhone-ceaster or Hrofe-Caestre, the camp of the Saxon warlord Hrof or Roffa, and a shorter step to the modern Rochester.

Try thinking about that during the annual summer Dickens Festival when Jingles, Pickwicks, David Copperfields, and a myriad of Dickensian characters roam the streets of Rochester. Or during the first week of May, when a unique Chimney Sweeps Festival takes place—a wild procession originating in the eighteenth century when sweeps flocked to the city to celebrate May Day. But a Dickens reference imposes itself on that ceremony too because of the author's valuable description of it in "Sketches by Boz."

It is a Kodak reference that imposes itself on Rochester in New York State, which derives its name from Colonel Nathaniel Rochester, who purchased in 1802 the land which became incorporated in 1834 as the City of Rochester. When George Head left to settle in Minnesota in 1855, he gave the name of his former hometown to the place that was to become the new state's county seat, famed for the Mayo Clinic.

In New Hampshire is another of the many Rochesters that exist in the United States. But it is the small and less well-known Rochester in Massachusetts, located near Wareham in the county of Plymouth and established as a town in 1686, that takes its name from the ancient city of Rochester in Kent, England, the place from whence many of the first planters emigrated.

Rockingham

An enchanting village in Northamptonshire, brimming with history, clings to the side of the hill that leads up to a castle. Seventeenth- and eighteenth-century stone houses with lichened slates line the steep main street of the unspoiled village of Rockingham with not much more than a post office and the Sondes Arms pub halfway up.

Its main attraction is Rockingham Castle, a fortress built by William the Conquerer, with two large round towers and arched gateway guarding the Welland Valley. The ancient stronghold has been the home of the Watson family since Henry VIII granted it to Edward Watson in 1530. He converted the former royal castle into a domestic residence, and extensions have been added ever since. The Elizabethan Great Hall with minstrel gallery contains a portrait of Elizabeth I. Paintings by Gainsborough, Van Dyck, Lely, and Morland hang on the walls of the long gallery, with its view from an end window into five counties. Dickens, a great friend of the family and a frequent guest, used the castle as the prototype for Chesney Wold in his *Bleak House*. He dedicated *David Copperfield* to the Watson family and wrote plays for them which were acted in the long gallery.

The thirteenth-century church of St. Leonard, considerably damaged in the 1640s during the Civil War and largely rebuilt in the nineteenth century, contains monuments of the Watsons. Like the castle, it overlooks and dominates the village from a dramatic setting at the top of the hill.

If a balance scale were set up for the two Rockinghams, with Old Rockingham Meeting House in Vermont on one side and Rockingham Castle in Northamptonshire on the other, the scale would tilt with the weight of years towards the castle that is nine hundred years old.

Rockingham in Vermont, settled in 1733, is known and much visited for the Old Meeting House, one of the finest examples of colonial church architecture remaining in New England.

When the town charter was granted in 1752, Governor Wentworth of New Hampshire chose to name it Rockingham in honor of the Marquis of Rockingham, Charles Watson Wentworth.

Rye

When the ancient sea town of Rye was designated a royal borough in 1289, it already had the distinctive title of Ancient Town with the status of a Cinque Port. Now it rises above the surrounding marshes, no longer a port. Its rivers silted up, the town has been left stranded inland nearly three miles from the English Channel. "Where now the sheep graze, mermaids were at play," wrote the poet Conrad Aiken. Within fourteenth-century walls are the cobbled lanes and half-timbered houses, many dating from the fifteenth century, which give the town its enormously appealing character.

The flat area known as the Salts lies below the imposing fourteenth-century Landgate, the proper entry into town from the north and the only remaining original gate of the three which once guarded the town. Camber Castle, built during the reign of Henry VIII as a defense against the French, is now a mile inland. The little fortress squats on the marsh between Rye and Winchelsea, while the two hill towns rise from the flat land which was once covered by the sea.

The impressively large Church of St. Mary crowns the hilltop of Rye and dominates the town. Below, the complex cluster of numerous old red-roofed houses presents a medieval picture with much to see in intriguing streets which invite aimless wandering down to the river level.

The church, largely rebuilt after the town was devastated by a French raid in 1377, features flying buttresses on the outside and a stained-glass window by Sir Edward Burne-Jones on the inside. A rewarding climb to the top offers a fine view as well as a look at the works of the famous old parish clock which strikes the quarters, not the hours. In the churchyard itself is an eighteenth-century brick Water House with a pleasing oval shape and tiled roof.

Just beyond the Church Square is Ypres Tower, built in 1249 as a defence against the French. Now a worthwhile museum of local history, it has an elevated terrace for another good view all around and down to the River Rother.

Two cobbled streets running out of Church Square are particularly rich in splendid old buildings—Watchbell Street, named for the alarm bell that warned when French were sighted in the Channel, and Mermaid Street.

In Mermaid Street, one of the most picturesque and famous streets in the world, is the equally well-known Mermaid Inn of 1420; the large medieval building became a coaching inn in the eighteenth century and a hangout for smugglers. The black-timbered architectural gem features oak-beamed interior, oak paneling, and a maze of twisting corridors. Near it is the fifteenth-century Hartshorn House, with three overhanging and distinctive gables. Known as "The Old Hospital" for having served as a hospital during the Napoleonic Wars, it had been the home of the Samuel Jeake family in the seventeenth century, given as dowry when the prosperous merchant (who built the wool storehouse opposite) married Elizabeth Hartshorn.

Jeake's House

181

The wool storehouse, known as Jeake's House, functions now as an excellent little bed and breakfast place. It was the residence of Conrad Aiken from 1924 to 1947. In more recent times, the novelist Rumer Godden resided at Number 4, near the top of the steep and delightfully picturesque Mermaid Street.

The Old Grammar School, named Peacock's School for the man who founded it in 1636, the former apothecary's shop with its lovely curved and bow windows, and the butcher's shop with its brass fittings—all may be singled out for special attention on the High Street.

Other streets reveal other treasures. Fifteenth-century Flushing Inn in Market Street features a sixteenth-century wall painting in the dining room. The Town Hall, with its arcaded ground floor, dates to 1743. Around the corner from it, a good view of the huge Perpendicular window of the church fills the end of Lion Street.

A seemingly inexhaustible list of attractions in Rye might render superfluous any need to mention the delightful towns and villages—Tenterden, Peasmarsh, Iden—and the great houses and castles—Hastings, Bodiam, Pevensey—which abound in the surrounding countryside.

Literary connections too are abundant. A sign on a building in Lion Street announces that John Fletcher, the Elizabethan dramatist and collaborator with Beaumont, was born in 1579 in Fletcher's House, now tearooms. Radclyffe Hall, whose *Well of Loneliness* created a sensation in 1928, lived in the High Street.

But the most popular literary figure is Henry James, who lived from 1889 to 1914 in eighteenth-century Lamb House. Afterwards, the novelist E.F. Benson, author of the *Mapp and Lucia* stories which are set in Rye, also lived in that superb Georgian house, now in the care of the National Trust and open to the public.

Although it seems unlikely that anyone would willingly leave such a paragon of a place, the story of Rye in New York State centers on two brothers who left in 1632. An infant colony on the shore of Peningo Neck began to be known by 1665 by the name of Rye to honor Thomas and Hachaliah Browne, the sons of Mr. Thomas Browne, a gentleman from Rye in the county of Sussex, England.

The Landgate

The name—a derivation of the French "rye" meaning "seashore"—applies equally to both. And both continue to rejoice in a common heritage. Rye mayors have exchanged visits. A stone from the walls of St. Mary's Church in the English Rye has been placed in the floor of Christ's Church Chapel in the American Rye. And the clock in St. Mary's Church was used as the model for the clock in the new city hall of Rye, New York.

How appropriate, if fanciful, it is to visualize the brothers Browne leaving their homeland through the ancient Landgate, which was adopted as a symbol on the masthead of the Rye *Chronicle* together with the words, "Ye Ancient Towne of Rye."

St. Albans

St. Albans in Vermont has as its namesake a city that is some two thousand years old. St. Albans in England is an attractive city about twenty miles north of London, known for its great cathedral, intriguing streets, old pubs and inns, markets, parks, and the remains of the Roman city of Verulamium.

The ancient city derives its name from the first Christian martyr in Britain. Alban, a Roman soldier stationed in Verulamium, was converted from paganism by a fugitive Christian priest whom he sheltered and helped to escape. For his new faith, Alban was condemned to death in the year 303. A church was built on the site of his martyrdom, and the shrine still stands in what has evolved to become, in 1877, a cathedral.

One good way to get the sense of a place is by viewing it, whenever possible, from a high perspective. In St. Albans, the panorama can be viewed from the centrally-located clock tower. This striking edifice (which also strikes the hours) was originally built as a curfew tower in about 1403. It houses a still older bell, cast in 1335, which can be seen from the winding stone staircase leading to the roof. The top affords a view over the city. Historic Watling Street was built by the Romans as a main route to connect the bustling city of Verulamium to another important Roman city, Londinium. On the western edge of St. Albans, Verulamium Park contains the ruins of that first-century Roman town and comprises most of the area of the old Roman city.

On market days, one may also witness the colorful scene beyond French Row, a charming street to the west of the tower, which imparts the feeling of a medieval town. Here were quartered the soldiers of the Dauphin of France sent for by the barons to force King John to adhere to Magna Carta. In one of the fourteenth century buildings, the Fleur de Lys Inn, King John of France is said to have been imprisoned after his capture at Poiters in 1354.

The Clock Tower

Nearby Holywell Hill, lined with Georgian houses, is named for the legendary Holly Well; King Arthur's father used its waters to heal the wounds he incurred in battle with the Saxons. The Waxhouse Gate, directly across High Street from the tower, is a covered passageway leading to the Cathedral and Abbey Church. In this archway, candles were sold to pilgrims on their way to the shrine of the saint.

The shrine inside the cathedral is a remarkably skillful restoration of the original, which was smashed to pieces during the reign of Henry VIII, at the time of the Dissolution of the

Monasteries. A nearby watching chamber guards the shrine. Only two such structures exist, the other being in Christ Church, Oxford.

The church itself has been restored, rebuilt, enlarged, and endlessly altered to result in the existing structure. The dates of construction are sobering. The pillars in the south transept, Saxon in origin, were incorporated into the present church by the Normans. The impressive Norman tower, the transept, the choir, the bays of the north side of the nave were built between 1077 and 1088 from Roman bricks taken from Verulamium down in the valley. The long nave, where massive piers and rounded arches are the chief architectural features, still gives an impression of what it must have looked like in Norman times. Colored murals and designs, dating from about 1215, were recently uncovered after centuries of hidden existence under whitewash.

Outside, the fourteenth-century Abbey Gatehouse, the single surviving monastic building, is the work of Abbot Thomas de la Mare, whose reign marks the high point of the monastery. He used the gatehouse as a prison, and subterranean dungeons have confined inmates from his time to the Napoleonic wars and beyond. Today, its inmates are students of St. Albans School, one of the oldest public schools in the country.

On the way down the hill to Verulamium, in Abbey Mill Lane, is the charming old inn, Ye Old Fighting Cocks, which claims to be one of the "oldest inhabited licensed houses in England." Used as the fishing lodge of the monastery, it later became, as the name implies, a cock-fighting center. Rebuilt in 1600 on medieval foundations, the pub suggests a history as interesting as its octagonal shape and makes a perfect stop for repast before continuing on to Verulamium.

Verulamium Park contains the ruins which mark the grandeur that was Rome. The initial view of the remnants of the original city wall is indeed impressive. Excavations have been ongoing, and a little imagination applied to the archeological remains can create a picture of the ancient planned Roman city with its forum, streets of houses and shops, two temples, two triumphal arches, and a unique theatre. The Roman theatre, not an

Ye Old Fighting Cocks Inn

amphitheatre, is the largest in Britain and the only one completely exposed to view.

A hypocaust, in relatively good condition, is a tribute to Roman ingenuity in coping with a need which still plagues inhabitants of Britain—a central heating system to contend with the weather. The suite of heated rooms, used for bathing, constituted the wing of a private town house; a large mosaic is featured in the warming room of the suite.

On the site of the Roman basilica where Alban was tried and sentenced before being led to execution is the Saxon Church of St. Michael, notable for its warmth of character and for its marble effigy of the city's illustrious resident, the statesman and essayist Sir Francis Bacon. There he sits in full life size, head resting on one hand, apparently fast asleep. The nearby museum, in the area where formerly stood shops and houses of the Roman forum, contains mosaics and various artifacts found in the area. Of particular interest among the Roman remains are a dolphin mosaic and the painted walls of one large town house.

Verulamium Park, by the River Ver, provides long, interesting walks and a site for contemplation. With the passage of time, the population gradually resettled in the area of the shrine. As Verulamium declined, St. Albans flourished and expanded to become a great city and not just the relic of a once-great city.

A final view of St. Albans from the valley below symbolizes the movement through the centuries. Look across the lake, formerly the monks' fishpond, at the upper portions of the cathedral. There is the imposing Norman tower, built of bricks which were once part of Verulamium.

And if imagination is allowed to run rampant, look across the ocean too, to the smaller, younger St. Albans which was inspired by this Hertfordshire city in Old England—to St. Albans in New England.

Sandwich

Once the most important port in England, Sandwich is now a small town in Kent on the River Stour some two miles inland from the sea.

The town was probably founded after the Romans departed from Britain in mid-fifth century, leaving behind a nearby fortification which is one of the most impressive Roman ruins in the country. Richborough Castle, or Rutupiae as they called it, was the first Roman base in Britain after the invasion by Claudius in the year 43.

Deriving its name from the Saxon "Sand-wyk"—the village or settlement on the sand—Sandwich became a great port, one of the original Cinque Ports. Of the original five ports, Hastings is now a prosperous seaside resort. New Romney is a somewhat decayed but pretty inland village. Hythe is another holiday resort. Dover has become, with the building of a great artificial harbor, one of the main ports on the English Channel. Only Sandwich remains a lively little market town, an enchanting place in this southeastern corner of England, which still conveys a sense of its past.

The purpose of the Cinque Ports was to supply the King of England with ships and men for fighting his wars—a kind of navy. In return, the ports received certain privileges. In effect, they were little sovereign states with their own rights and courts and with freedom from taxation by the Crown.

The Cinque Ports confederation was formed for mutual support in the reign of Edward the Confessor (1042-1066). Rye and Winchelsea were designated "Ancient Towns" and added to the confederation to alleviate the strain of providing for the needs of the king. Towns which joined later were known as "Limbs." Sandwich was most important, and royalty frequently stayed there.

The wars with France kept Sandwich Haven (as the port was called) busy. Here ships gathered between 1203 and 1216 to try to recapture the lands lost to France by King John. The victorious Battle of Sandwich was fought in 1217 and is still remembered by the Hospital of St. Bartholomew, built to commemorate the victory. Another crucial conflict took place in 1293 when the Cinque Ports demolished the French fleet. The fifteenth century saw a reversal of fortune with several severe attacks by the French. In 1457, an expedition from Honfleur inflicted serious damage and resulted in the deaths of so many citizens including the Mayor, that the Mayor of Sandwich still wears a black gown as the official robe.

The supremely important Haven of Sandwich began to deteriorate by 1500 because of silting up of the mouth of the river. Indeed, nature was winning a war against all of the ports, and by the time Henry VIII came to the throne, the days of greatness of the Cinque Ports and their subsidiaries were over. Henry VIII, anxious about the plight of the port, visited Sandwich in 1532 and again in 1539. Elizabeth I visited in 1573 and was petitioned to improve the harbor. However, nothing could be done to save it. Silting was leading inexorably to the end of Sandwich Haven.

Eventually, in 1890, the land which covered the former flourishing port became one of the finest seaside golf courses anywhere. Today, only Dover is a functioning port, and it alone retains a link to former maritime glory.

A certain Cape Cod town retains a link to glorious old Sandwich. In 1634, a shipload of 102 persons, including twenty-four adults and twenty-one children from Sandwich, embarked at the ancient port for the American plantations of New England. Five years later, the General Court in Boston allowed the original Indian name of the Cape Cod settlement to be replaced with the name of Sandwich.

Another linking is interesting, if irrelevant. A descendant of the first Earl of Sandwich, a compulsive gambler, took his meals in the form of some bit of food inserted between two slices of bread in order not to interrupt his game. Although he had no connection to the town, Sandwich was immortalized in a word now known throughout the world.

The Fisher Gate

The ancient town has itself been given a kind of immortality. The whole of the old town within the original defenses has been designated a conservation area, thus ensuring that the exceptional number of valuable and attractive timbered buildings will be preserved. The old part is the same area as that within the medieval walls; and the population of five thousand is not significantly higher than it was in Elizabethan times.

Earth ramparts follow the line of the old wall and provide a pleasant tree-lined walk around much of the town. The maze of narrow streets and alleys within is filled with a jumble of old houses and cottages—a mixture of medieval, Tudor, and Georgian buildings with paneled walls, oak beams, or plaster-work ceilings. Many have cellars which once had the legitimate purpose of storing wine and later the illegitimate purpose of hiding smuggled goods.

In Strand Street are the best of the timbered houses. Near St. Mary's is the King's House, where Henry VIII and Queen Elizabeth I lodged. The datestone of 1713 on the brick front is deceptive, for the brickwork of that date conceals a timber-framed house of about 1400. The thirteenth-century Long House, fifty-two feet in length, was once known as Herring House, for this was the place where herring dues were paid. The Pilgrims and the Weavers are fine examples of fifteenth-century houses. At the end of Upper Strand Street is a splendid house named the Salutation, designed in Wren style by Sir Edwin Lutyens.

The riverside Strand Street leads to a bridge and ancient barbican before opening onto the quay. The Barbican Gate of 1539, with twin turrets, collects tolls from traffic crossing the bridge over the River Stour. The Fisher Gate on the quay is even older—1384. Nearby, is the old Customs House, marred by an eighteenth-century brick facade covering the medieval timber-framed house.

In the heart of Sandwich, in the Market Square, is the Guildhall of 1579 with an oak-paneled court room on the ground floor. The council chamber above is hung with portraits of such notables as Edward Montagu, a great naval hero and the first Earl of Sandwich who paid the town a compliment by adopting

Norman Tower of St. Clement's

its name for his title. The Guildhall also maintains a small but rich collection of objects of local history including original charters and documents.

St. Peter's, in the center of Sandwich near the Guildhall, is one of three surviving medieval churches. The clock on its tower can be seen from almost every part of town. The curfew bell has rung from St. Peter's every evening at eight since the thirteenth century. But formerly it proclaimed the time for releasing hogs on to the streets to scavenge. Nowadays, although the job of refuse disposal is done by dustmen, the bell nevertheless continues to ring.

St. Mary's was left in derelict condition when it was damaged by an earthquake in 1578 and a collapsed tower in 1667. But it has recently been restored, and some Norman work remains in the west end.

The parish church of St. Clement's is the largest and most important of the three. Its pure Norman tower, perhaps the finest in England, massive and elaborately embellished with three rows of arcading, dates to 1100. Inside are richly carved choir stalls, a magnificent fifteenth-century font adorned with heraldic shields, and monuments of the prosperous members of the medieval community. In the chancel is an unusual feature—a medieval system of sound amplification consisting of holes in the stone below the choir stalls and a similar set of holes high up on the sanctuary walls.

Sandwich has no stupendous monuments. But the meanderer is easily rewarded at nearly every step. Quaint old houses are everywhere. In King Street, the Dutch House is an excellent example of houses built in the style of their homeland by Dutch refugees who fled from Spanish oppression in the 1560s, bringing their weaving skills with them and adding to the prosperity of the town.

A plaque on a small house in New Street marks the place where Thomas Paine lived in 1759. He is proudly remembered for the single year he spent in Sandwich before he left for America and published his revolutionary writings.

Outside the town wall just off the Dover Road is the medieval Hospital of St. Bartholomew, an ancient almshouse with a thirteenth-century chapel, erected to commemorate the Battle of Sandwich fought in 1217 on St. Bartholomew's Day, the 24th of August. The hospital founded in honor of the saint still functions as a charity for the needy. Tradition is strong. Each Saint's Day, the practice continues of distributing Bartlemas biscuits stamped with the hospital seal and founding year of 1190; in fact, food and lodging were provided here for pilgrims as far back as that original date.

Sandwich today ought to be slowly relished for the endless tidbits it offers—green walks along the ramparts, an ancient fresh water channel made in 1287 known as the Delf flowing under ancient houses, narrow and intriguing lanes. . . the story of some thirteen centuries. Countless bits unite to make a savoury Sandwich.

Happily, the edible pun can be carried to an extreme in England: Sandwich in Kent can be coupled with Brown Bread Street in Sussex, With Ham in Wiltshire, and Beer in Devon.

Shrewsbury

Shrewsbury, the county seat of Shropshire, is a picturesque black-and-white town in which timber-framed Elizabethan buildings dominate. A strategic location only twelve miles from the Welsh border has influenced its history and architecture. Offa's Dyke, built by King Offa in the eighth century to mark the boundary between England and Wales, is still visible. The thirteenth-century castle, rebuilt by Edward I during campaigns against Wales, now contains a regimental museum. All around the central square, with its old Market House of 1596, narrow medieval streets filled with half-timbered houses make a splendid background for envisioning the Shakespearean dramatization of an episode of the bloody Battle of Shrewsbury of 1403 when Falstaff tells the future King Henry V that he himself killed Percy Hotspur after fighting "a long hour by Shrewsbury clock."

Two bridges, the English Bridge on the east and the Welsh Bridge on the west, cross the River Severn. The town, enclosed within the loop made by the river, began to prosper and develop after the Norman Conquest. It became increasingly important, particularly in the wool trade. The Market Hall in the square was built in 1596 to accommodate the sale of Welsh wool in the upper room, while the lower level was for the sale of dairy products and vegetables. One gate of the thirteenth-century town wall still survives, St. Mary's Water Gate. Today, ten bridges cross the Severn at Shrewsbury, but the present English Bridge built in 1775 and the Welsh Bridge of 1795 are still the main accesses to the hub, although the modern town has spread well beyond the confines of the river.

The town center is small, no more than half a mile by half a mile, making perambulation pleasurable in this museum of architecture with Tudor houses, elegant Queen Anne and Georgian buildings, and fine Victorian structures such as the railway station. Curving streets, curious alleyways called "shuts," and mysterious street names such as Wyle Cop add appeal for the meanderer.

St. Mary's Church has the third tallest spire in England and a fine collection of medieval stained glass. St. Alkmund's and St. Julian's stand close by.

Situated in the town center, within the horseshoe bend of the river, is the early seventeenth-century Lion Inn at the top of Wyle Cop, one of the oldest streets of medieval Shrewsbury. Books on the changing face of Shrewsbury, with the familiar format of before and after pictures, omit the Lion because its appearance has hardly changed. The attractive hotel with a lion over the entrance can boast of such guests as Thomas DeQuincy (author of *The Confessions of an English Opium-Eater*), King William IV, Paganini, Jenny Lind, and Benjamin Disraeli. The most famous lodger, however, Charles Dickens, has been honored by a suite named for him.

In America, the town of Shrewsbury in Massachusetts has the honor of being named for the delightful Shropshire town across the sea.

Springfield

The parish of Springfield, as seen today, presents itself as a tiny but attractive village or suburb of Chelmsford.

Oliver Goldsmith is believed to have lived in Springfield for a time while he wrote his "Deserted Village." Although the grounds are thin that Springfield is actually the prototype for that poem, it can be viewed as a source of inspiration.

The heart of any English village is its church, and the Church of All Saints, largely Norman, has a delightful setting among tall trees beside a village green. Among the many items of interest inside the church are its unusual fourteenth-century window tracery, mid-thirteenth-century carved font, medieval

Church of All Saints

stained glass, memorial brass of 1421 of a man in armor, a fifteenth-century chancel screen, and a Tudor funeral helm. Of particular historical interest to Americans is an ancient tablet listing church wardens. Among the names inscribed is that of William Pynchon, a resident of Springfield who sailed with the Pilgrim Fathers.

One of the original incorporators of Massachusetts Bay, Pynchon led a group of emigrants from Roxbury in 1636 to the wilderness of the west banks of the Connecticut River. His settlement was the first in the western part of what was to become Massachusetts. Within a few years, the settlers secured official status as a town, and the name of Springfield replaced the original Indian name of Agawam to honor the English home of their founder.

Although Pynchon was successful as a trader, he ran into trouble with a book he wrote on his religious beliefs, "The Meritorious Price of Our Redemption." Printed in London, the book, which expressed views contrary to official dogma, was condemned and publicly burned in Boston. Thus, he may have been the first to be banned in Boston. Dishonored and removed from his position as judge at Springfield, he returned to England in 1652.

Despite his unhappy experience in the early history of the New World, William Pynchon is remembered as the founder of the now great industrial city of Springfield in Massachusetts, a city with stands in complete contrast to the quiet, hardly noticeable parish of Springfield in Essex.

Sudbury

The old Saxon town of Suthburgh, the fortified town in the south, has retained the essence of its ancient name and character. The date of its actual founding is not known, but Sudbury is first mentioned in the Anglo-Saxon Chronicle of 797. In that era, the town was surrounded by a moat which is marked in modern Sudbury by Friars Street, one of the attractive winding streets which leads out from the market square.

Sudbury has had a weekly market since Saxon days, as mentioned in the Domesday Book of 1086. The town is still vibrant with activity on Thursdays, when open stalls in the open space of Market Hill purvey a variety of produce and goods. Although Sudbury has a modern shopping precinct which attracts people from all around, it is Market Hill that is magnetic, particularly on market days.

In this focal point of Sudbury is a statue of the town's most illustrious citizen and one of England's greatest painters, Thomas Gainsborough. The house in Sepulchre Street in which he was born in 1727 is now open to the public as a museum; and the street, renamed Gainsborough Street, runs most appropriately into Market Hill, where the bronze statue of the beloved painter, palette in hand, stands in front of the west door of St. Peter's Church and overlooks the bustling scene.

St. Peter's has become a redundant church. Built in the fifteenth century, it is being used for concerts and public functions. All around the open square are other noteworthy buildings. On the south side is the public library, formerly the Corn Exchange. To the north of the church is the grey brick Town Hall of 1828. An attractive Georgian house contains Lloyds Bank.

Interesting old buildings exist everywhere in Sudbury. The Chantry is a timbered and gabled house on Stour Street with a finely carved corner post. Next to it is Salter's Hall, the home of a fifteenth-century merchant and the most elaborate of the timbered houses surviving in Sudbury. Also of the fifteenth century is the old Moot Hall, with a pretty oriel window and overhanging story. A fifteenth-century gate is all that remains of the Dominican Priory built in 1272 and dismantled at the Dissolution of the Monasteries by Henry VIII. The Ship and Star is a four-hundred-year-old inn which may have been a pilgrim's guest house for the neighboring priory.

Statue of Gainsborough in Market Hill

These ancient timbered houses and fine old churches are evidence of the wealthy wool trade which once characterized Sudbury. Keeping the past while adapting to the future, Sudbury still carries on an important weaving industry, especially famous for silk.

St. Gregory's, the mother church of Sudbury, stands in a quiet area at the west end of town. It was built by Simon of Sudbury on the site of a wooden seventh-century church. Who was Simon of Sudbury? The Archbishop of Canterbury. He was beheaded by the Wat Tyler rebels during the Peasants' Revolt in 1381. His head was displayed on London Bridge for six days and then sent back to Sudbury where it is still displayed in a case in the vestry. In a less grim vein, the church contains a medieval font cover of ingenious design. Highly ornamented, it rises in stages to a height of about twelve feet. The lowest stage can be pushed up in telescopic fashion, allowing the font to be used without disturbing the upper part.

All Saints Church, mostly of the fifteenth century, is on the lower part of Friars Street near Ballingdon Bridge. When the churches of Sudbury were rebuilt in the flourishing fifteenth century, the fourteenth-century chancel of All Saints was allowed to remain, and it remains, consequently, the oldest building in Sudbury.

The town has literary associations, for it is the "Eatanswill" of Dickens' *Pickwick Papers.* The Swan Hotel was known as Buff Inn in that novel, and the Rose and Crown (now a cinema and shops) was the Town Arms. John Bunyan, the Puritan writer and author of *Pilgrim's Progress*, reputedly visited Sudbury, staying at the home of his friends the Burkitts in what is now Burkitt Lane.

And it has American associations, for Sudbury was the home of a key figure in the American Revolution. William Dawes arrived in New England in 1635 and eventually settled in Boston. His house in Sudbury Street remained in the possession of the family for five generations, until it was pulled down by the British in 1775 during their occupation of Boston. It was William Dawes, born in Boston in 1745, who made the famous

202

ride with Paul Revere to "spread the alarm through every Middlesex village and farm. . . ."

Earlier, in 1630, a Puritan lecturer from Sudbury, John Wilson, sailed under John Winthrop of Groton to New England. John Winthrop noted in his diary the death through sickness on that voyage of "Jeff Ruggle of Sudbury and divers others of that town." But John Wilson survived to become a leading director of the colony at Charleston and founder of its church.

Edmund Brown, born in nearby Lavenham, had served in the Sudbury, England, church for fourteen years. He became the first minister of Sudbury, Massachusetts, and was probably responsible for conferring on a New England wilderness settlement the name of Sudbury as his gesture in carrying a bit of local heritage to the New Canaan.

East Anglia was a Puritan stronghold and a principal source of emigration. But Sudbury sent more emigrants to New England than any other town in East Anglia. Perhaps that is why Sudburians on both sides of the Atlantic are surrounded by such familiar town names as Haverhill, Needham, Ipswich, Boxford, Newton, Acton, and Braintree.

Taunton

In Somerset's fertile Vale of Taunton Deane, in the midst of an area of exceptional beauty and venerable history, lies the county town of Taunton. All around are open fields, hills covered with gorse and heather, small villages with ancient churches, and manor houses and cottages, mostly of local red sandstone. Stately homes and castles, as well as abbey ruins and apple orchards abound. Natural beauty is happily balanced with light industry as the Taunton Cider Company continues to supply the nectar of the west from the fruited lands. The impressive past displays one prehistoric hill fort south of Taunton called Castle Neroche and another hill fort in the nearby village of Norton Fitzwarren.

The Quantock Hills, to the north, is a poet's place. William Wordsworth and his sister Dorothy came in 1797 to live in the Quantocks at Alfoxton House; and Coleridge lived just three miles away at Nether Stowey in a cottage (now open to the public) in which he wrote "The Ancient Mariner"; nearby Watchet is the small supposed port from which the eponymous character embarked.

When poets have not lived here, they have been inspired by it. The seventeenth-century poet Michael Drayton composed lines extolling its bounty:

> What eare so empty is, that hath not heard the sound
> of Taunton's fruitful Deane, not matched by any ground?

And Defoe described Taunton itself in 1702 as a "large, wealthy, and exceedingly populous town."

Taunton was founded by King Ine of Wessex who defeated the the British in 710 and built a castle for the defense of his realm. But only twelve years later, his fortress was destroyed. The town prospered in Norman times under the protection of the

Bishops of Winchester, and a massive castle established in 1138 has been a focal point of Taunton history.

A gateway to the west of England, Taunton has endured endless comings and goings and has been involved in tumultuous events. In 1497, an insurgent army of Cornishmen, marching one hundred fifty miles away to London to protest taxation, were defeated at Blackheath. In the same year, Perkin Warbeck, claiming to be the second son of Edward IV, one of the little princes murdered in the Tower, led his rebellious army through Taunton on his way to London to seize the crown from Henry VII. The pretender met his demise at Taunton and was forced to stand trial in the Great Hall of the castle.

Taunton was involved in the Civil War of the 1640s and changed hands several times. In 1642 it was captured for Parliament, and the following year it was taken for the King. In 1645 Taunton was under siege with Admiral Robert Blake, the parliamentarian leader, fighting and defending the town against the king's forces. At the Restoration, Charles II deprived the town of its charter and dismantled the castle. But a second charter was granted in 1677, and the ruin was rescued in the eighteenth century. The east gatehouse is now incorporated into a hotel, and the gateway is the entrance to a museum.

During the Monmouth Rebellion, the Duke of Monmouth passed through Taunton where he was proclaimed king of England. But after defeat at the Battle of Sedgemoor in 1685, retribution was severe. Judge Jeffreys tried the supporters of the rebellion at the Bloody Assizes of 1685, and 509 were condemned to death while hundreds more were deported to the West Indies for life. The trials took place in the castle, still the center of Taunton.

From the wide, busy shopping street, the narrow Castle Bow passes under a portcullis to the Castle Green. This archway was the main entrance to the castle in medieval times. In modern times, new is imposed on old. The Castle Green, no longer green, is a parking area. In the midst of all that antiquity, a modern sign informs: "Ladies Toilet in Castle Walk 130 yards to the right."

The market, formerly held in Castle Green, has been moved to another part of the town where the sale of livestock, produce

Castle Bow

and general goods continues to take place on Saturdays. Taunton has kept its position as an important market center for over a thousand years.

The castle houses the Somerset Museum. A contemporary portrait by Kneller of Lord Chief Justice Jeffreys hangs in the Great Hall where once the voice of the judge bellowed out his sentences and where his ghost reputedly still walks on September nights (the month of the Bloody Assizes). On display are relics of local history covering a wide range and expanse of time from Stone Age antiquities, ancient pottery, and a prehistoric canoe, to musical instruments and costumes, a seventeenth-century silver collection, and a Van Dyke portrait of King Charles I and his queen.

On another side of the Castle Green is the Castle Hotel. With walls that date back to about 1300, it offers architectural interest as well as culinary and lodging comforts. The nearby Norman Garden contains part of the walls of the inner moat of 1160 and a square Norman well. Just beyond is the River Tone, which runs down from the Brendon Hills in the west to flow through the center of Taunton and give the town its name.

While the castle is at the heart of Taunton, two church towers dominate the skyline. The imposing parish church of St. Mary Magdalene was built toward the end of the fifteenth century when Taunton had developed a thriving wool trade. A magnificent tower of Ham Hill stone, completed in 1514, made the church perhaps the finest example of Somerset architecture. The soaring tower of 163 feet (rebuilt in exact replica in 1862 to avoid possible collapse) enhances and completes the perfect vista from eighteenth-century Hammett Street, an approach of unsurpassed Georgian elegance. The interior of the church features the fan-vaulted ceiling of the tower and a beautifully carved black oak roof. Fifteenth-century glass glitters from the great clerestory windows.

Also handsome is the Church of St. James. Probably in existence in 1127, it has been so altered and rebuilt that little remains of the original structure, and the present building dates from the fourteenth and early fifteenth centuries. Its red sandstone tower, the outstanding feature of the exterior, rises to a height of 120 feet. Inside, ancient timbers support a barrel roof, an unusual feature in Somerset churches. A fifteenth-century font is carved with Christian figures, and a Jacobean pulpit is carved with non-Christian mermaids.

Two sets of almshouses belong to the seventeenth century. Huish's Almshouses in Magdalene Street were founded in 1615 by a London merchant named Richard Huish, while Gray's Almshouses were founded in 1635 by Robert Gray of London, who was born at Taunton.

Of further interest in the town center is the Tudor House, an impressive example of medieval domestic architecture with gables and half-timbered construction, and with parts of it dating to the fourteenth century. If it is true that Judge Jeffreys dined here during his Bloody Assizes, it is appropriate that the oldest house in Taunton is now a restaurant.

To the dramatic events of the seventeenth century must be added a coda—the establishment of Taunton in America.

Cohannet was purchased in 1630 by settlers who thanked God for bringing them "over the great ocean into this wilderness from our dear & native land. . . and in honor and love to our dear and native country," their report continues, "we called this place Taunton."

The American city of Taunton has outgrown the mother town in size, population, and industry. But the old, formerly-tumultuous town of Taunton has earned its tranquility.

Tewkesbury

A town steeped in history, Tewkesbury was the scene of a major battle in 1471, the Yorkist victory in the Wars of the Roses. It is an old town of timbered, black-and-white houses and good inns situated in Gloucestershire where two rivers, the Avon and the Severn, meet.

An old watermill on the Mill Avon River, now a restaurant, is in a picturesque spot off Church Street opposite the abbey church. Parts of the thirteenth-century bridge which crossed the Avon are incorporated into the present King John's Bridge, with its four quaint triangular pedestrian "refuges" on either side. But the Mythe Bridge over the Severn, designed by Thomas Telford in 1826, is Tewkesbury's most famous bridge. Today the rivers provide recreation with boat trips from the center, sailing and fishing, and a marina. But in the past, river commerce brought prosperity to medieval Tewkesbury, as did the wool trade, and buildings throughout the town recall that important place of an earlier era, a town influential enough to possess one of the seven copies of Magna Carta.

The most obvious sign of former prosperity and the most impressive structure in town is Tewkesbury Abbey, which dominates from its position at the western end of town. Completed in 1121 with imported Caen stone, the prodigious church retains its Norman character although it has undergone several restorations. The church is the oldest building in town and all that remains of the abbey founded in the reign of William Rufus, son of William the Conqueror. The second largest parish church in England, its Norman tower is one of the largest and best examples in existence. Measuring 46 feet square and 148 feet high, the tower, with its pinnacles and graceful interlaced arches, rises over Tudor houses and reveals a view from its top over the Severn and Avon valleys to the mountains of Wales.

The basic simplicity of the cruciform exterior is repeated in the beautiful interior with its many outstanding architectural features: nave arcades supported on huge round pillars, vaulted ceilings dating to 1340, roof bosses in the nave, a series of chapels against the east wall, and medieval stained-glass windows of the fourteenth century in the chancel. One intriguing tomb depicts the decaying corpse of Abbot Wakeman with five creatures—worm, snake, frog, mouse, spider—crawling over it. There is also a lifelike effigy of Edward Despencer, a knight who died in 1375. The old abbey survived the Dissolution of the Monasteries when the townsfolk purchased the glorious treasure from Henry VIII in 1539 for £453.

Church Street, High Street, and Barton Street, the three main streets in town, form a "Y" shape which has remained the basis of the town's development. A hundred timber-framed buildings from the sixteenth and seventeenth centuries stand in the three major streets. Two particularly good ones in the High Street are the Old Fleece and the Ancient Grudge.

In the town center are a number of architecturally fascinating inns. The Bell Hotel opposite the abbey in Church Street, probably the most photographed, has a thirteenth-century wall painting in the dining room. It was here that Mrs. Craik wrote her fourth novel, *John Halifax, Gentleman,* published in 1857. The Black Bear is mainly early sixteenth century, but tradition dates it to 1308, making it reputedly the oldest in Gloucestershire. The sixteenth-century Tudor House has a priest's hole in the chimney of the present coffee room. Charles Dickens gave fame to the attractive Royal Hop Pole Inn by referring to it in *The Pickwick Papers.*

Also in the High Street is the Swan Hotel with the jettied House of the Nodding Gables next to it. A notable pair of half-timbered houses in Barton Street have been converted to form a museum and tourist information center, perhaps the best place from which to begin an exploration of the town.

A unique survival in Church Street is the range of tiny fifteenth-century shops known as Abbey Cottages. These primitive homes were daily transformed into shops by letting down the window shutters to serve as counters. One unit remains as a reconstructed example, while a Georgian house stands incongruously in the middle of the row, and the John Moore Museum stands at the end to commemorate the local novelist whose *Portrait of Elmbury* (1945) is set in Tewkesbury between the two world wars.

The town is fortunate in having preserved an unusual market. Tewkesbury claims to celebrate the largest and most ancient mop fair in the country. Originally a hiring fair when workers and servants would carry mops and tools of their trade, it is now an annual funfair that takes place in October.

A tourist may, with a bit of lucky timing, witness a re-enactment produced by residents from time to time of the famous battle in which the Lancastrians were decisively defeated on the field still called the Bloody Meadow. Indeed fortunate is the tourist who is able to sojourn to the Gloucestershire town of Tewkesbury.

Bypassed by the Industrial Revolution and by the railway age, Tewkesbury has managed to retain much of its medieval character. Not totally bypassed by the twentieth century, it attracts only light industry and population, located as it is near a motorway network. The town has come a long and pleasant way since those ancient times when it was known as Theocsbury, when a hermit monk named Theoc supposedly built his hut in the vicinity.

Toppesfield

A Saxon chieftain named Topa or Toppa is undoubtedly the source for the name of this village, which has been variously called Toppesfend, Toppesford, and Thopefield. It is *not* the topmost village in the county but is located in a rather flat agricultural area in the northern part of Essex about sixteen miles beyond Braintree and fifty miles from London.

In 1905, the Reverend H. B. Barnes, a former Rector of St. Margaret's Church, wrote a description of the village in a "Sketch of Toppesfield Parish, Essex Co., England" in which he decried the fact that the town was struggling in the midst of an agricultural depression. He deplored the neglected state of a village being abandoned by the young who leave it in search of employment elsewhere. The situation seems to be still applicable today.

Away from the main road, it is a desolate village which seems to be suffering still from depression, agricultural or otherwise, which entices the young to move elsewhere. Even the pump in the town center seems lonely. And the church is locked when not in use.

Through the attractive lych gate is the church yard from which one sees the rather unattractively-faced walls of the early Tudor church building, enclosed by surrounding council houses and school. Only the village hall opposite the church seems to be a beehive of activity. And perhaps the conspicuous red telephone booth in the center is another source of activity.

Reverend Barnes described the poor condition of the church:

> The body of the church has nothing to recommend it, the seats are mean looking and uncomfortable to use, the pulpit is commonplace, the west gallery. . . is faced on its pillars with carved oak; the great oak beams which span the nave are similarly cased, and unhappily neither they nor the roof are in a sound condition.

Town Pump

Nevertheless, the church of St. Margaret offers an impressive distant view of its handsome brick tower with four unusual and elaborate corner pinnacles, dated 1699. Inside the church, mainly of the early sixteenth century, several brasses in the chancel and a thirteenth-century Purbeck marble figure of a cross-legged knight in armor are now unfortunately hidden beneath the organ. A fragment of fifteenth-century stained glass is in one of the Perpendicular windows, and well-preserved parish registers date back to 1558.

The names of Samuel Symonds and his wife Dorothy can be found in the register as well as the baptism records of their ten children, born between 1621 and 1633. Samuel Symonds later retired to New England, where he was granted a farm of five hundred acres, partly within the bounds of what later became Topsfield. It was he who brought about the change of name to the area settled in 1639 as an offshoot of Agawam, as Ipswich was first known. In 1650 the site received its new name—

213

Topsfield—for Symonds remembered with gratification his former home, even if he did not remember the correct spelling.

"What is its future to be?" the Reverend Barnes asked of Toppesfield and proceeded to make this prediction:

> Automobilism, or electric railways, will make travelling easy, and then this corner of Essex, with its healthy climate, its quiet beauty, its fertile soil, its fine oaks and other trees will attract the class of persons who want a nice house and a few acres of land. Then land will again fetch in this district ten times what it fetches now; then there will be plenty of employment in stables, gardens and pleasure farms for the men who now flock into towns. But this will not be in my day.

Church of St. Margaret

214

Nor has his prediction of an influx of wealth come to pass in our day. Nevertheless, fulfillment is beginning to be apparent; old houses are being renovated, new ones are being constructed. The forecast may be a trifle premature, the changeover slow; but the growth and development of an area rich with resources is inevitable. A sense of the past is leading to ascent of the future. Already, the Reverend's *Sketch* is, happily, in need of updating and revision.

Truro

The one-hundred-year-old city of Truro is nearly one thousand years old. While the *town* of Truro received a charter from Henry II in 1156, it was declared a *city* only as recently as 1877. The contrast of old and new is characteristic of Truro today.

The cathedral, with its three spires, rises above the encroaching, narrow streets and slate-roofed houses and dominates this capital city of Cornwall. Completed in 1910, it was the first cathedral to have been built in England since London's St. Paul. When the decision was made to restore to Cornwall a diocese of its own, a diocese which was lost over eight hundred years earlier when it was merged with Exeter, Truro was chosen for its central position in the county.

Not the style of the cathedral, but its materials are Cornish—granite primarily. Although only just over a century old, it is reminiscent of great medieval cathedrals with its Gothic style of architecture. In design and in spirit, the cathedral links the present to the past. The line of the nave bends at a slight angle to the chancel to accommodate the existing street.

Built on the site of the sixteenth-century parish church of St. Mary, it incorporated most of that ancient south aisle into the new building, giving a warmth and character that is normally lacking in a totally new and sterile construction. St. Mary's aisle still has some ancient stained glass, a seventeenth-century alms box, and monumental brass. The organ and pulpit are of the eighteenth century, while brasses and monuments date from 1567 and parish registers from 1597. In the Jesus Chapel of the cathedral is an altar painting of Christ blessing Cornish industry, a new treatment of an old subject which appears on many walls of old Cornish churches.

Crooked streets in the area between the cathedral and the river offer the flavor of the distant past when Truro was an important river port. Now another kind of flavor comes from a small bakery in the shadow of the cathedral which makes two distinctively Cornish biscuits known as Cornish gingerbreads and fairings.

Truro Cathedral

The town's wealth emanated from its location at the center of a rich tin-mining district. Truro was a medieval "coinage" town. By order of King John, tons of tin ore would be brought here from the mines to be tested and stamped. A corner (*coin* in French) of each block of tin was cut off for testing by officers who would certify that the quality met set standards. By the end of the sixteenth century, the assaying and coining of a third of the Cornwall tin was done in Truro. Its prosperity is evidenced by the charter granted by Elizabeth in 1587 for two weekly markets and three fairs.

The shipping of tin and copper ore from Truro was another source of wealth. But the town suffered a serious decline when it lost its trade to the rival port of Falmouth in the seventeenth century. Like Falmouth, Truro is located on the River Fal, but further inland, away from the hazards—and the conveniences—of the sea.

The fortunes of Truro revived in the eighteenth and nineteenth centuries, and the town's present charm derives from the prosperity of this period. Wealthy merchants built their magnificent Georgian mansions, and a few—the Boscawens and the Lemons—gave their names to its streets. The town's Georgian past is still to be seen in the gracious Lemon Street, in the cobbled Boscawen Street, and in High Cross.

The name of Truro itself has had many variations in both spelling and meaning. Orthographic variations include Triverv, Triueru, Truueru, Treuru, Truru, Trurow, and Truroe. As for etymology, some believe that the root "tri" suggests that originally Truro was made up of three streets, an accurate description in the first centuries of its existence, from about 1160 to 1460. Or it may have meant the "town or castle on the river." In any case, the origin of the name has disappeared together with its Norman castle.

Truro is a commercial center, unrivalled west of Plymouth. In the last century, its mining and mercantile business and the wealth of its gentry gave the town the epithet of Little London. The compliment is returned for there is a lot of Cornwall in London, where many of the bridges crossing the Thames are made of Cornish granite.

Truro is also the cultural center of Cornwall with an Art Gallery that has paintings by Rubens, Hogarth, Gainsborough, and Constable. Yet, for a city, Truro is comparatively small, with a population of just over twelve thousand. A brief walk from the city center will take one into the Cornish countryside. The wide valley filled with villages that are quaint and picturesque in both appearance and name contains, for example, just two miles south of Truro, Come-to-Good, derived from *cum ty coit*, or the valley by the cottage in the wood. The village offers peace and simplicity in the whitewashed, open-timbered roof of its Quaker Meeting House of 1710.

A city of contrasts, Truro may also be contrasted with the American Truro. Not the Cape Cod seaside holiday atmosphere, but a commercial, inland, bustling metropolis describes this capital city of Cornwall.

Waltham Abbey

Waltham must be a good name for a town for there are duplications and variations of it on both sides of the Atlantic. In England, there are Walthams in several counties, including Great Waltham and Little Waltham, Waltham-on-the-Wolds, and Waltham Cross (which takes its name from the cross erected by Edward I, one of twelve such crosses, in memory of his beloved Queen Eleanor). In New England, there is a Waltham in Maine and one in Vermont. But the Waltham in Massachusetts, on the Charles River, incorporated in 1738, is the counterpart of the ancient town of Waltham Abbey in Essex, on the River Lea.

Waltham Holy Cross was the less popular official name for the pleasant town of Waltham Abbey in England. Located sixteen miles from London, the busy market town was founded by Tovy, standard bearer of King Canute, who brought the "Holy Cross" to this weald-ham or forest homestead.

Tovy made an important abbey town of the existing Saxon settlement of huts in the clearing by the river. He built a church in about 1040 to accommodate the relic and some three dozen people. But growth was inevitable because of the arrival of pilgrims from all over to the site of the true cross. Thus, while the origins of this ancient town may be rooted in the legend of a miracle, it might have ended as a legend had not the cross brought about the miraculous growth of a great abbey town.

Tovy's church was rebuilt in about 1060 by King Harold who is said to have prayed here before the Battle of Hastings, perhaps not well enough, for his body was brought back for burial after the battle against the invading Normans. A stone in the Waltham Abbey churchyard marks the position where King Harold is believed to be buried. After their conquest, the Normans could not realistically be expected to give the full allotment of respect to the dead king, the last Saxon monarch, and facts are therefore difficult to substantiate. Nevertheless, the fact that Waltham

Abbey Church is affectionately known as Harold's Church, would seem to be reason enough to bring the name of Waltham to America.

Why not recall Harold's Church, with its inspiring and emotional associations, in the new homeland? William Brown, Samuel Livermore, Daniel Benjamin, and others petitioned to have their precinct of Watertown become a separate township

Waltham Abbey Church

with the name of Waltham. Perhaps the new homeland resembled that former home on the River Lea. But it now has the excesses of modern city life—population, industry, and traffic. The City of Waltham in Massachusetts today stands in direct contrast to the relatively quiet and ancient town of Waltham Abbey.

Although the Dissolution of the Monasteries, following the break of Henry VIII with the Pope, reduced the Abbey Church of the Holy Cross to a third of its former size, it remains the highlight and glory of the town and the oldest Norman church in England. Henry II had altered the abbey considerably in 1177 when he founded an important Augustinian priory as he sought expiation for the murder of Thomas à Becket. Seven years later, the priory became Waltham Abbey. English royalty arrived here regularly from the time of Harold to Charles II. The town took on aristocratic associations and was granted charters for fairs and a market. It grew.

The nine-hundred-year-old abbey is a magnificent ruin. The monastic part, with central tower and transepts, was destroyed in the Reformation; but the nave, having been used as a parish church, is preserved. The solid Norman interior has massive stone piers, some carved with chevron and spiral decoration, to support the arcading and upper clerestory. The east end of the nave has fine stained-glass windows of 1861 by Burne-Jones above the altar and a restored ceiling with signs of the zodiac. The Lady Chapel on the south side of the nave, dating to 1316, survives with a restored fourteenth-century wall painting of the Last Judgment. Remnants of herringbone work on its outside wall are believed to belong to the early church founded by Tovy and incorporated into Harold's Church some twenty years later.

Also surviving in a picturesque part of the town are the bridge and Abbey Gatehouse of 1370. With its wide and narrow entrances for carriages and pedestrians, the gatehouse is a fine fragment of the ancient monastery.

The Market Place of Waltham Abbey is close to the church. Here the weekly Tuesday market, first granted by Henry II in the twelfth century, is still held in the open space. Also in the Market Square is the fifteenth-century Welsh Harp, an oak-framed inn dating back to those early times when pilgrims traveling in large numbers to the shrine of King Harold needed accommodation. With lych gate into the courtyard, the inn is believed to have been the guest house of the monastery.

Waltham Abbey was formerly a forest town exclusively for the pleasure of kings. No building was permitted within the Royal Forest of Waltham, and cruel penalties were inflicted on anyone who in any way interfered with forest laws and royal pleasures. Safer to kill a man than a deer. So the town had a favorable position as an accessible forest town for sovereigns who resided at the Tower in London, from William of Normandy to Elizabeth I. Henry VIII is said to have been particularly fond of it, and legend has it that he went to the hunt on the morning of Queen Anne Boleyn's execution.

Gradually, the forest ceased to be a playground for princes, and the town was to develop merely as a market center for Lea Valley produce and cattle. The forest was threatened with being built in or enclosed, and it was abused. It was cut down and its timber used to build ships. But in true English tradition, it was rescued from possible demise. Now called Epping Forest, no shooting is permitted in what was once a hunting forest. Although it is still the home of deer and other wild life, it is primarily the happily-hunted ground for tame picnickers escaping from London, whose property it became. The Corporation of the City of London acquired Epping Forest, and Queen Victoria dedicated it to the public in 1871.

Waltham tells a compelling story, and many have been inspired by it. Thomas Tallis, the Elizabethan composer, was organist of the abbey at the time of the Dissolution in 1540. John Foxe, author of the famous *Book of Martyrs* lived here. And Tennyson, who lived for a time in nearby High Beech, composed "The Bells" based on the sounds he heard from the bells of Waltham Abbey Church. An appealing if fictional account relates an inspirational tale of another variety: Queen Elizabeth's Hunting Lodge at Chingford, now a museum, was supposedly the royal hunting house in which James I carved a joint of beef with his sword and dubbed it "Sir Loin."

223

The town still retains its ancient appearance in the midst of twentieth-century life, and Harold's Church is still the center of it all. Waltham Abbey illustrates the progress from a cluster of huts to an appealing modern town in just over nine centuries.

Wareham

Small and quaint old Wareham, with its lovely buildings and picturesque quay by the bridge, is known to tourists who pass through. That's the trouble: visitors merely pass through and miss the valuable and pleasurable points of interest.

Situated between two rivers which flow eastwards into Poole Harbor, the Frome and the Piddle, the Dorset town of Wareham, with its obviously defendable setting, was inhabited over two thousand years ago. The ancient Britons were followed by the Romans. Then the Saxons made it a place of great importance and fought hard to defend Wareham from the invading Danes. Much of the town was destroyed, including its castle and nunnery, before the Danish occupation in 866. After King Alfred liberated it from the Danes, Wareham continued to develop and became an important town with two mints and a wealthy port. Now the harbor is silted up and Wareham lies a mile inland.

The extremely ancient town walls still enclose Wareham on the north, east, and west; the south side was protected by the River Frome. The walls were originally ancient British earthworks before the Romans reinforced them with stones and retained them as valuable strategic defenses. Today the earthen walls which surround the town offer fine views of Wareham and glimpses into the past. In the northwest corner, a curved section is believed to have been a Roman amphitheatre. The highest section of the wall is known as Bloody Bank because some of the Monmouth rebels sentenced by Judge Jeffreys were executed on that spot in 1685.

Of the three old church buildings remaining in Wareham, Holy Trinity, with its sixteenth-century west tower and fourteenth-century nave, is now disused and has become an arts center.

The embattled tower of Lady St. Mary church rises behind the buildings at the east end of the quay. Adjoining it was the Benedictine nunnery founded by St. Aldhelm in about the year

225

700 and later destroyed. The present church has survived for over a thousand years as one of the largest Saxon churches in the land. Although spoiled by the excesses of the Victorians when they rebuilt it in about 1842, it contains a number of worthwhile objects.

In the north aisle is the stone coffin of Edward the Martyr, King of Wessex and England, who was murdered in nearby Corfe Castle. His body was buried in Wareham in 978 but later removed to Shaftesbury. Short stone pillars are believed to be remains of pagan altars, and a cresset stone of the Middle Ages has little hollowed-out cups which held oil for burning floating wicks. A unique nine-hundred-year-old, six-sided lead font and two thirteenth-century Purbeck marble effigies of cross-legged knights in armor, both of the thirteenth century, are also in the parish church of Lady St. Mary.

St. Martin's Church

A more modern effigy in the Church of St. Martin is that of T. E. Lawrence—Lawrence of Arabia—who died in 1935. Sculptured by Eric Kennington, the recumbent figure is represented in Arab costume, with curved dagger in hand and head resting on a camel saddle.

St. Martin's Church, high on the northern walls, was founded by St. Aldhelm in about 700. Although Saxon parts still remain, the chancel arch and north aisle are of the very early Norman period, and most of the building dates from the twelfth and thirteenth centuries. Medieval paintings cover walls of this, the most interesting of the churches in Wareham.

The town center was largely destroyed in 1762 in a burning reminiscent of London's great fire of 1666. Much of the present town was rebuilt in the simple brick style exemplified in the Red Lion Hotel.

Wareham is arranged in grid street pattern, probably of late Saxon origin, and the central crossroads cut Wareham into four quarters. The Red Lion is in the western arm or West Street. In the eastern arm, East Street, is located an interesting set of almshouses of 1741. North Street leads up to St. Martin's at the northern entrance of the town, and South Street leads to the river and to Holy Trinity at the southern entrance. In South Street are two of the town's best houses. The Manor House, built of Purbeck stone, with three stories and top balustrade, is an imposing example of eighteenth-century domestic architecture at its best. Nearly opposite is the Black Bear Hotel, a striking inn of about 1800, with a lovely facade and a columned porch carrying, naturally, a black bear.

Located 117 miles from London, Wareham is the gateway to the Isle of Purbeck, a puzzle of a name that refers, not to an island, but to the peninsula which extends to the English Channel in the south and west and to Poole Bay in the east. Wareham and its surroundings contain historic as well as literary associations worth persuing.

T. E. Lawrence lived seven miles northwest of Wareham in a former gamekeeper's cottage called Clouds Hill, a name as flamboyant as the person himself. He wrote *Seven Pillars of Wisdom* in the house now owned by the National Trust and open

227

to the public. Killed in a motorcycle accident in 1935, Lawrence was buried in the churchyard in Moreton village, where the small church has chancel windows engraved by Laurence Whistler.

Wareham is Anglebury in Thomas Hardy's Wessex. The stretch of moorlands northwards towards Dorchester and eastwards almost to Poole, is Egdon Heath in *The Return of the Native* and in other works by Hardy.

The ruined and romantic Corfe Castle, only four miles from Wareham, offers superb views. Its name means "gap" for it is situated in a break in the Purbeck Hills where the River Corfe and a tributary have isolated the site on which the castle stands.

Corfe Castle

A gap of another kind can be happily closed—the gap in the experience of those trippers on their way to holiday places such as Swanage and Bournmouth who rush past the small, sleepy town.

Wareham in Massachusetts was unquestionably named for the Wareham in Dorset, England. The similarity extends beyond the name to the streams of vacationers who also avoid the newer Wareham as they hurry on to Cape Cod.

Weymouth

Weymouth is a Georgian town associated with George III, the first king to stay in a seaside resort. His bathing experiment must have been successful, for he returned many times to the town that honored him with a major landmark, a statue on the Esplanade commemorating the king's jubilee in 1810.

The Esplanade, also noted for Queen Victoria's Jubilee Clock, follows the sweep of the glorious bay, unrivalled on the south coast. A five-mile stretch of sandy beaches and safe bathing justifies the town's popularity as a resort. Pleasant terrace houses of the eighteenth and early nineteenth centuries, some with fine ironwork balconies, some with bow windows or canopies, face the sea along the extended frontage from the harbor to the end of the promenade.

Weymouth, with its well-sheltered harbor entrance, has been a port since Roman times. Now ferry services from Weymouth cross the English Channel to Cherbourg and to the Channel Islands.

Situated near the quay is the old town with its winding streets, aged inns, chandlers' shops, and seafaring ambience. Primarily a seaside resort, Weymouth is not rich in buildings of great architectural interest, but a few may be singled out. The Gloucester Hotel was once a favorite residence of George III, and two Tudor cottages in Trinity Street are now a museum displaying typical Elizabethan furnishings.

Many small and unspoiled villages are within easy reach. A four-mile walk eastwards to Osmington Mills offers reward in the form of freshly cooked lobsters from inns overlooking the sea. High on a hillside there is also the Osmington White Horse, one of a number of white horse figures in the southern part of England, including the pre-Roman white horse at Uffington. Formed by scraping away the encroaching grass from the chalk surface, the huge equestrian figure at Osmington, created in 1808, is unique. It is the largest, measuring 323 feet high and

280 feet long; and it has a rider, supposedly George III. Some believe that Wellington is the man who rides his high horse, but in *The Trumpet Major,* Thomas Hardy described the "huge picture of the king on horseback" being cut by workers "removing the dark sod so as to lay bare the chalk beneath."

Weymouth is Budmouth in Hardy's Wessex, and the author refers to it as the "royal watering-place." Made popular by the king's presence when he resided at Gloucester Lodge (now the Gloucester Hotel), the Georgian watering-place, Hardy writes, "would be nothing if it wasn't for the Royal Family, and the lords and ladies, and the regiments of soldiers, and the frigates, and the King's messengers, and the actors and actresses, and the games that go on."

A beautiful situation near the sea is the reason also for the existence of the American Weymouth, the oldest English settlement in Massachusetts after Plymouth. Early traders were attracted to the shores of Wessaguscus, also known as Wessagusset, by its possibilities as a port. Thomas Weston, a London merchant, gathered a company of about sixty men who landed in 1622 inside the bay later known as Boston Harbor to develop a trading post. But the ill-fated Weston colony, with its unmotivated and disorganized band of men unwilling to undergo labor and privation, was destined to fail. Another colony was started in the autumn of 1623 under charter from the Plymouth Company, and settlers arrived from Weymouth in England in following years.

In 1635, the General Court of the Massachusetts Bay Colony allowed the Reverend Joseph Hull to settle at Wessaguscus with his company of faithful followers consisting of twenty-one families from Weymouth, England. The colony thrived with the influx of about one hundred respectable and highly motivated persons. In September of that year it was incorporated and the name changed to Weymouth in memory of the pleasant port in Dorset they had so recently left.

Winchester

The massive statue of Alfred the Great, erected in 1901 to mark the thousandth anniversary of his death, dominates the High Street of Winchester; and that is entirely appropriate, for the Saxon King Alfred made Winchester the dominant city of England—the capital, in fact. And even when the capital shifted by the time of Henry III to London, sixty miles to the northeast, kings continued to come, and Winchester continued to be important. Today it is a city resplendent in the riches of history and tradition that make for an unforgettable visit.

Long before King Alfred, the site was an important Belgic settlement in the Itchen valley where the River Itchen goes through a ridge in the chalk hills. Under the Romans, it was given the Latin name of Venta Belgarum—Venta of the Belgae. The Romans, with their strong concern for security, made this tribal center a major town of the area, and Venta became the fifth largest city in Britain. After the beginning of the fifth century, the Saxons occupied the place referred to as Vintanceastir by the Venerable Bede, who traced the ancient history of England to the year 731.

Because the Normans traditionally preserved existing sites of worship, the cathedral overlaps the site of the earlier Saxon church of King Alfred. On low ground, the cathedral is overlooked by St. Catherine's Hill to the south and by St. Giles's Hill to the east, which gives the best panoramic view over the city and the cathedral.

Begun in 1079 and almost continuously rebuilt or remodelled up to the fifteenth century, the cathedral supplies architectural examples ranging from the Early Norman (with the most complete work in the transepts) to the Late Gothic.

Soon after its initial completion in 1200, the eastern part behind the altar had to be enlarged to accommodate pilgrims coming to the Shrine of St. Swithun. The legendary saint is held responsible, unjustifiably, for England's not-so-legendary rainy

High Street with King Alfred's Statue

weather. The humble Swithun, who died in 862, had requested burial in the churchyard. When monks afterwards tried to remove the body to a more exalted position, a deluge rendered the task impossible, and rain continued for forty days. Amateur meteorologists still maintain that rain on the 15th of July, St. Swithun's Day, will continue for forty days.

The undramatic exterior of the cathedral belies the architectural splendors of the interior. Entering from the west, one is immediately impressed by the massive fourteenth-century pillars from which shafts tower up and soar into the intricate fan vaulting. The resultant feeling of height and grace can be attributed to the remodelling work done under William of

232

Wykeham (1366-1404), who made it the longest cathedral in England, measuring 556 feet.

In addition to the spacious and unmatched nave, the priceless treasures of the interior include the original fifteenth-century altar screen, seven exquisitely carved chantry chapels, medieval wall paintings, a twelfth-century black Tournai marble font, the library containing rare and beautiful illuminated manuscripts, and reminders of the thirty-five monarchs who dwelt in this ancient capital.

An entire day could be spent profitably inspecting monuments and inscriptions inside the cathedral. Almost as many kings and queens lie buried in this cathedral as in Westminster Abbey. Stone screens in the chancel carry six carved chests that contain the bones of Saxon monarchs. The remains of William II were brought here when he met a violent and mysterious death while hunting in the easily accessible New Forest. He was buried under the tower with, the records say, "many looking on and few grieving."

In the north aisle is the grave of Jane Austen, and on the wall above is a memorial tablet and window. Perhaps the most remarkable thing about the inscription is that it makes no mention of Jane Austen the revered author whose books have brought joy to so many, but refers to Jane Austen the beloved human being: "The benevolence of her heart, the sweetness of her temper, and the extraordinary endowments of her mind obtained the regard of all who knew her and the warmest love of her intimate connections."

In a small chapel of the south transept, the famous angler Izaak Walton lies buried. A memorial window was given by fishermen of England and America in 1914. He is still remembered for his best seller of 1653, to give it the full and fashionably long title, "The Compleat Angler, Or the Contemplative Man's Recreation, Being a Discourse on Fish and Fishing, Not Unworth the Perusal of Most Anglers."

One somewhat amusing epitaph is outside near the path leading to the west front of the cathedral. The inscription on the gravestone of Thomas Thetcher, who died in 1761 "by drinking Small Beer when hot," admonishes:

233

Soldiers be wise from his untimely fall.
And when ye're hot, drink strong or none at all.

Also outside, in the secluded cathedral close, a harmonious blending of various buildings includes a thirteenth-century deanery, a half-timbered Tudor building called Cheyne Court, and the Pilgrim's Hall which was a lodging place in the Middle Ages for pilgrims on their way to Canterbury.

Beside the cathedral close is one of the city's two remaining gates, the fourteenth-century Kingsgate with the tiny Church of St. Swithun above it. Clearly, Anthony Trollope had this building in mind when he described the Church of St. Cuthbert at Barchester:

> It is a singular little Gothic building, perched over a gateway, through which the Close is entered, and is approached by a flight of stone steps which lead down under the archway of the gate. It is no bigger than an ordinary room. . . but still it is a perfect church.

Kingsgate leads to College Street, one of the prettiest streets in Winchester. Number 8, a stucco-fronted building looking towards the cathedral, is the house in which Jane Austen died. Also in College Street is the school founded by Bishop William of Wykeham in 1382 to equip boys for entry into New College, Oxford, which he had already established. Winchester College, the oldest public school in the country, is nearly opposite the palace of the bishops of Winchester. And within the palace grounds are the ruins of the twelfth-century castle built by Henry de Blois.

Of the former Norman castle, which was the royal residence, only the hall survives. Inside, a circular board seventeen feet in diameter ostensibly attests to the veracity of the Arthurian legends. Like the legends, King Arthur's Round Table makes a good story, but the facts are highly debatable.

Castle Hall stands near Westgate, the other of the two remaining gates and a grand monument to the times when

Winchester was a walled city. From the top of Westgate, the room which houses a small museum gives access to an excellent view of Winchester's High Street, selected by many as *the* street in England with the greatest number of historical associations. Between Westgate at the top of the High Street and the bronze statue of King Arthur at the other end, are a great many features worth exploring: streets with descriptive names include Staple Gardens (formerly the "staple" where wool was marketed) and Jewry Street (the former Jewish quarter). The Royal Oak Inn of 1630 has a subterranean bar and a long, well-worn table that has

Godbegot House

been in use for several hundred years. From the turret of the old Guildhall (now Lloyds Bank), the curfew bell still rings each evening at eight as it has done since the time of William the Conqueror. Godbegot House, with a name of doubtful origin, is an attractive timber-framed Tudor building.

A focal point of the High Street is the early fifteenth-century Butter Cross, the place for the sale of butter and eggs in the past and for Sunday newspapers now. Several lovely old houses are very near the Cross, and just beyond is a covered shopping way or colonnade called the Pentice.

Where the High Street broadens out into the Broadway, beyond the current Victorian Guildhall and St. John's Hospital, is the huge and impressive statue of Alfred the Great; he stands in splendid dignity, some eighteen feet high, dressed in Saxon helmet and mantle.

The High Street ends with a stone bridge over the River Itchen, beside which is a watermill built in 1774, now a youth hostel. Further along, at the foot of St. Giles's Hill is the early sixteenth-century Chesil Rectory.

Any tour of Winchester should include what is possibly the finest set of medieval almshouses in the country—the Hospital of St. Cross. Just over a mile from the city center is the foundation established in 1137 by Henry de Blois, Bishop of Winchester and grandson of William the Conqueror. It still provides homes for thirteen poor men, who can be identified by traditional garb: loose gowns, flat hats, and silver crosses.

Still another tradition provides for the distribution of the wayfarer's dole of bread and ale to travelers who apply for it at the Porter's Lodge. Perhaps no longer home-baked or home-brewed, perhaps only a symbolic portion, the tradition nevertheless offers clear evidence of a need for sentiment and stability in this city of memories—the same need which prompted emigrants to choose old and familiar English names for their new homes in America.

Woburn

Old Woburn, which dates from Saxon times, was the site for the founding of a great Cistercian abbey in 1145. The village grew steadily up to Tudor times, then declined. But with the coming of the stage coach, *the* method of travel in the eighteenth century, it revived as a coaching town and was largely rebuilt in distinctively handsome Georgian style. The coming of the railways brought an end to coaching and to the revitalization of Woburn. As seen today, it remains the elegant Georgian village it once was.

An attractive High Street preserves many fine buildings. The Bedford Arms Hotel is a dignified eighteenth-century coaching inn and main staging post on the road from London to the North. It was rebuilt after a fire of 1724 destroyed most of the town. But the signpost of the inviting Bolyn Tea Rooms boasts of its "XVI Century" origins. For another kind of drink, pubs are plentiful. So are attractive shops. The bay-fronted butcher's shop, formerly the Goat Inn, is particularly pleasing. Tudor buildings with eighteenth-century facades house antiques and boutiques and encourage the tourist trade. And nearby Woburn Abbey, with thousands upon thousands of visitors each year, has insured the twentieth-century revival of the town of Woburn.

Henry VIII gave Woburn Abbey to the Russell family, Dukes of Bedford, after the monastery was dissolved. But not until the seventeenth century did the Russells come to Woburn to live. In 1626, when the plague was raging in London, the family fled to their home in the country to escape infection. They liked it and decided to rebuild the abbey as a family residence. Remodelled and reconstructed several times since then, the present house, built around a quadrangle on the monastic site, dates from 1746.

The stately home stands in the middle of a park of some three thousand acres in which eleven varieties of deer roam freely. The house was passed through succeeding generations to the Duke who inherited the abbey along with £5,500,000 worth of

Woburn High Street

death duties. Crippling taxation forced him to open the house to the public in 1953, and now visitors may feast on the priceless collections of art treasures contained within.

Paintings include works by Van Dyck, Rembrandt, Reynolds, Gainsborough, Velazquez, and Holbein. The dining room is called the Caneletto Room because it contains twenty-one views of Venice commissioned when the painter Caneletto came to England in 1746.

State apartments have gilded ceilings, French and English eighteenth-century furniture, and a vast variety of art treasures. Especially richly furnished is Queen Victoria's State Bedroom, as it became known after Queen Victoria visited with Prince Albert in 1841.

The China Room displays dark blue and gilt Sevres dinner service, the most complete set of this quality in existence anywhere. It was presented by Louis XV in 1763 when the Duke was the English ambassador at the signing of the Treaty of Paris.

A curious grotto exemplifies shell rooms which were so popular in the eighteenth century. Thousands of shells set in stucco are arranged in patterns to decorate walls and ceilings of a room which was originally open to the garden.

Great entrepreneurial skill has come to the rescue. The estimated cost of over a thousand pounds a day for the upkeep of Woburn Abbey is met by the entrance fees of the throngs that arrive for a pleasurable day's outing. A Wild Animal Kingdom allows visitors to drive through the safari park among lions and

Woburn Abbey

tigers. A crafts center, an antiques center with about fifty shops, a passenger train, and restaurants are all designed to obtain revenue needed to preserve the sumptuous setting for posterity.

Not in the Wild Animal Kingdom of Woburn in England, but in the wilderness of seventeenth-century New England, was there a need to preserve the name of the place in Bedfordshire which produced emigrants to Massachusetts. Captain Edward Johnson wanted to honor his friend, Major-General Robert Sedgwick, who came from Woburn in 1635. The General Court met in Boston in 1642 and ordered that "Charlestowne Village is called Wooborne."

Woodbridge

Woodbridge is a Georgian market town with an old Anglo-Saxon name—Udebryge. Established in 920, the Suffolk town near the coast of East Anglia became a fairly busy port early in its history. With a location on the estuary of the River Deben, which opens out to the North Sea, the town has inevitably taken on seafaring associations.

The early population gradually moved from the banks of the Deben to higher ground, and the town developed around the area known as Market Hill. The area around the quay, the most picturesque part of Woodbridge, remains pleasantly crowded with boats and yachts; and a special attraction, the eighteenth-century tidemill, has been preserved and is available to the public for viewing. Another attraction is a mile away on the opposite bank of the Deben, where the Saxon burial remains of Sutton Hoo were excavated in 1939 to reveal an Anglo-Saxon ship with the huge and magnificent treasure hoard of an East Anglian king of the early seventh century.

The walk up to Market Hill takes in many fine old Tudor, Jacobean, and Georgian buildings. A thriving sixteenth-century trade in boat building and in the production of sailcloth, rope, and other maritime provisions is the basis for the prosperity currently reflected in the rich merchants' houses in town. Also of interest, at the Bell and Steelyard Inn in New Street, is a unique seventeenth-century overhanging structure for hoisting up and weighing the loads of hay and corn of waggons entering town.

On Market Hill, the Elizabethan Shire Hall is perhaps the most striking building. Completed in 1570, its lower floor was open for use as a market. Bricked up in Regency times, it became a corn exchange before being converted to its current use as a court room. The grand Elizabethan structure, with Tudor and Georgian homes and shops all around, was built (like so many other treasured local buildings) by the town's benefactor,

Thomas Seckford. He lived just outside the town at Seckford Hall, which functions now as an excellent little hotel.

Also in the Market Square, the King's Head Inn, half-timbered and overhanging, is another notable example of Tudor architecture. Carved corbel faces on the Seckford Street side are supposedly heads of the original innkeeper and his family.

The fifteenth-century Parish Church of St. Mary, with a tower 108 feet high, rises at the top of the hill to dominate the town from its position overlooking rooftops, countryside, and river. Its flint and stone flush-work is typical of Suffolk churches. The spacious interior contains a monument in the north chancel to Thomas Seckford, the town's Elizabethan benefactor, who also endowed the almshouses on a hillside site in Seckford Street, rebuilt in 1869 and recently modernized.

A more recent Woodbridge personality is Edward Fitzgerald, the poet and translator of *The Rubaiyat* of Omar Khayyam. Born at nearby Bredfield House, he is associated also with the Bull Hotel, on the lower side of Market Square, where he regularly put up friends and visitors, including Alfred, Lord Tennyson.

The true glory of the town, the tidal River Deben, is indirectly responsible for the town's name. "Woodbridge" probably derived from a wooden bridge (over a brook flowing through a valley and into the Deben) which led to the settlement on Market Hill. Other towns named Woodbridge have probably received their names in a more arbitrary manner.

Woodstock

The area in New York around Woodstock was part of the the hunting grounds of the Esopus Indians when European emigrants first began to arrive and to settle. By about 1762, the settlement was finally established under Robert Livingston, who had a sawmill on the site. Eventually, he gave the permanent name of Woodstock to his settlement. The reason is not known for his conferring that name on what was to become the most famous of the American Woodstocks, especially since the founding of the Woodstock Music Festival gave the Catskill town a measure of notoriety in the sixties.

Several other American Woodstocks were already in existence, including those in Vermont and New Hampshire. In New Hampshire, it is believed, the name was chosen to honor the young and rich Viscount of Woodstock.

But the first of the new world Woodstocks was originally part of Massachusetts until shifting borders placed it in Connecticut. It is no accident that Judge Samuel Sewall chose the name in 1690, recording in his diary that he bestowed it in honor of Woodstock in Oxfordshire. Curiously, the first American Woodstock is, like its counterpart in England, located not very far from Oxford.

The English Cotswold stone town, situated some six miles northwest of Oxford, was in ancient times part of the forest of Wychwood. The name is a corruption of "Woodstoe" or the "woody place." An important hunting site, it was recorded in the Domesday Book as a royal manor.

Woodstock had been a favorite hunting forest for kings from the time of Alfred the Great. In the twelfth century, Henry I built a seven-mile stone wall around the deer-stocked forest and created the first enclosed park in England. The town grew up as a royal borough at the edge of the park to service the needs of royalty.

Henry II resided here with his mistress, the Fair Rosamund. In 1163-4, he gave plots of land for the building of hostelries to accommodate his men. And he established the Tuesday market. The Black Prince was born here in 1330. In 1453, Henry VI granted the town's first charter. Henry VIII often stayed with Catherine of Aragon. His daughter Elizabeth was imprisoned here during the reign of Mary. Royalty frequented the lodge at Woodstock up to the time of James I.

The town became famous in the sixteenth century for its local industry of glove manufacturing. Deerskin from Wychwood Forest usually supplied the raw material which was cut out by men and then delivered to women and girls to be made up. Elizabeth I is among the distinguished visitors to have been presented with Woodstock gloves. Shops still purvey them, and Woodstock remains one of the last centers for the industry.

The English Woodstock is today a small town of about two thousand people with attractive stone buildings, gift and antique shops, and pubs and hotels catering to the influx of tourists.

Its center extends from the triangular market place to the main gates of Blenheim. Near the market place, the parish church of St. Mary Magdalen retains its Norman south doorway, fifteenth-century west porch, and eighteenth-century tower but has otherwise been almost entirely rebuilt in 1878. Opposite is Fletcher's House, a sixteenth-century former merchant's home that serves now as the headquarters of the Oxfordshire County Museum and displays the history, archeology, and culture of the county from Neolithic times to the present. A Town Hall built in Palladian style in 1766 has fine Assembly Rooms hidden behind the sixteenth-century Marlborough Hotel. The Marlborough Arms, the Bear Hotel (dating in part from the sixteenth century), and the Feathers are among the many welcoming inns in town.

But the chief glory of Woodstock today, and the reason for the influx of hordes of tourists, is Blenheim Palace, built by John Churchill, the first Duke of Marlborough. When Queen Anne rewarded him with the ancient manor of Woodstock for his victory over the French in the Battle of Blenheim in 1704, he

244

commissioned Sir John Vanbrugh to design the enormous mansion covering seven acres.

The stately house sits in an equally magnificent park of two thousand acres landscaped by Capability Brown in the 1760s. He replaced the geometrical layout with informal groupings, created a lake above and below the bridge by damming the River Glyme, and extended the lawn up to the palace—resulting in one of the finest examples of naturalistic eighteenth century park landscape.

The magnificent masterpiece of golden stone contains collections of furniture, paintings, sculpture, and tapestries in gilded state rooms. Open to view are such rooms as the Long Library with ten thousand volumes, the Great Hall embellished with Grinling Gibbons stone carvings, and the bedroom in which Winston Churchill, grandson of the 7th Duke of Marlborough, was born on 30 November 1874. Sir Winston is buried in a contrasting and movingly simple grave in the Bladon churchyard at the southern edge of the park.

The sophisticated town of Woodstock presents a prosperous face to the world. Located on the Oxford to Stratford-upon-Avon road, it is well placed to continue to attract tourists from all over to the town at the gates of Blenheim as well as to the home of the Dukes of Marlborough.

Worcester

Ask an older, loyal Worcester native about his city and he is almost certain to recall, not without anger or bitterness, some wonderful house or other which was pulled down "for no other reason than that it was standing." So spoke the curator of the Edward Elgar Birthplace Museum. To an outsider, any city than can boast of Elgar and music has enough. But Worcester, despite the mutilation wreaked by so-called city planners, has much more.

Worcester today is a city known for its music and its cathedral, as well as for its china and sauce. It is a modern city with traffic, factories, offices, architectural insults, and noise— all conspiring to conceal ancient buildings and history.

Worcester must have been much gentler, much quieter, very long ago when it was merely an important ford on the River Severn in pre-Roman times. Its importance rose in Saxon times, with the establishment of a bishopric in the year 680 marking the first major historical event. Under the Saxons, when it was known as Wigorna ceastre, the town grew. The Normans were wise enough to retain the Saxon Bishop Wulfstan after their conquest, and Worcester continued to grow under his leadership and inspiration. He rebuilt the cathedral in 1084. The crypt of that cathedral remains unaltered and is one of the finest examples of Norman work in the country with its dramatic use of columns and bays.

The chief glory—some would say the only glory—of Worcester is the cathedral. From any distant point of view the cathedral is imposing. It stands on a high bank above a bend in the River Severn, displaying its majestic and massive central tower. Across the Severn, on the west bank, the County Cricket Ground affords one of the spectacular views of the cathedral.

From close up, the site and surroundings are equally impressive. Ancient out-buildings which formerly belonged to the cathedral extend to the river. On the south side of the

246

The Cathedral Crypt

cathedral, a Norman gateway leads into the cloisters. Beyond the cloisters and the adjoining decagonal Chapter House of about 1140 is College Green, a quiet refuge where one can quickly forget the noisy streets and view the monastic buildings and ruins. Houses in the Green include the Old Palace in Deansway, ruins of Guesten Hall of the monastery, and the refectory, which became the King's School. Founded by Henry VIII in 1541 after the monastery was suppressed, the King's School provides the boys' voices of the cathedral choir.

Inside, design and style blend agreeably. The nave was constructed over a period of two centuries, from the mid-twelfth to 1377. The two westernmost bays of the nave date to about 1160. An unbroken view of the high vaulted roof with pointed arches on pillars stretches for a distance of over four hundred feet to give the satisfying feeling of harmony and proportion.

The choir, begun in the thirteenth century, contains fourteenth-century choir stalls with one of the finest sets of misericords in England, their carvings representing ordinary life in medieval times. In the center of the choir is the tomb of King John, who died in 1216. He had requested burial in the cathedral of his favorite city, and his effigy in Purbeck marble is the earliest royal sculptured figure in England.

Another royal memorial is Prince Arthur's Chantry, built by Henry VII in memory of his eldest son, who died of pneumonia in 1502 at the age of fourteen. It is a suitable spot for reflection on what the course of history might otherwise have been. After the premature death of Prince Arthur, his young widow, Katherine of Aragon, became the first of the wives of Arthur's younger brother, Henry VIII.

Other buildings of architectural merit exist in various parts of the city. Friar Street contains the finest group of early timbered houses and gives an impression of the medieval city which prospered as a cloth-weaving center.

The Greyfriars, originally the guest house of the friary, was built in about 1480. Saved from demolition, it has been lovingly restored and is now maintained by the National Trust.

On the opposite side of the street is the Tudor House Museum, a sixteenth-century timber-framed building with molded plaster ceiling and exposed example of the original wattle-and-daub construction. The museum displays life in Worcester from Elizabethan times.

Two houses in the city have royal associations. Through the back door of the King Charles House in New Street, the young Prince, later Charles II, escaped after his disastrous defeat at the Battle of Worcester, with Cromwell's troops in hot pursuit.

On the corner of Queen Street and the Trinity stands Queen Elizabeth's House, again a half-timbered house. She visited in 1574, and tradition has it that she addressed the crowds from its open gallery.

Worcester seems to have made little provision for pedestrians. If one can safely cross the road to get there, at the end of Friar Street, in Sidbury, is the Commandery, perhaps the most famous of the city's black-and-white buildings and the only museum in Britain devoted to the English Civil War. Its Great Hall with hammerbeam roof, oriel windows, and minstrels' gallery is most impressive. A carved oak Elizabethan staircase leads to an upper room with a sixteenth-century wall painting. Originally founded as the Hospital of St. Wulfstan in 1085, its odd name is believed to derive from the title taken by masters of the hospital from about the thirteenth century. More likely, the current name emanates from its use by Prince Charles as a base during the Battle of Worcester.

The Berkeley Hospital, one of the loveliest groups of alms-houses in England, was founded in 1682 by Robert Berkeley, grandson of Judge Berkeley who left six thousand pounds for that purpose. His statue stands at the far end of what appears to be a pretty Netherlandish picture by Peter de Hooch.

Another building to admire is the Guildhall in the pedestrianized High Street. Rebuilt in 1721, it has statues of Charles I and Charles II in niches on either side of the entrance, while a carving depicts Cromwell's head nailed by its ears above the doorway; in a niche between two upper center windows is a statue of Queen Anne garbed in brocaded dress.

From Worcester, Prince Charles managed to escape. From Worcester, the Severn escapes. And Worcester offers musical escape too. Sir Edward Elgar was born in 1857 in Upper Broadheath, just three miles outside the city. His cottage, with its inspiring view of the Malvern Hills, has become the Elgar Birthplace Museum and contains many items associated with the composer who made an immeasurable contribution to English music. His father kept a music shop in Worcester at 10 High Street, now occupied by the department store of Russell and Dorrell.

The tribute to music does not stop there. Worcester is known for its Three Choirs Festival which it shares with Hereford and Gloucester. Held in turn each year in each of the three cities, the festival claims to be the oldest in existence. It traces its origins to the early eighteenth century when it began as a charity to help orphans and widows of the clergy. It continued to evolve from 1715 as London artists began to supplement the provincial choirs. Elgar's music is well represented in festival performances, and

249

Elgar himself is commemorated by a window in the cathedral with the theme of "The Dream of Gerontius."

The Elgar Birthplace Museum

In addition to this cultural feast, Worcester is world famous for at least two products useful in a culinary feast. The chemical experiments of Dr. John Wall (who died in 1776) resulted in a formula for the improved manufacture of china and led to the present Worcester Royal Porcelain Company. The Dyson Perrins Museum contains a collection of porcelain from the beginnings to the present day.

Worcestershire Sauce dates from the early nineteenth century when Lord Sandys, a former governor of Bengal, brought a recipe back to England. He asked the chemists Lea & Perrins to prepare the recipe. They consented. But the finished product was found to be so unpalatable that they—because the English never throw anything away—relegated the jars to their cellars.

250

A fortunate accident occurred when they came across those sauce jars some years later and decided to taste it again before discarding the disagreeable stuff. The sauce, having matured, was absolutely superlative!

It is perhaps a more idealistic venture that binds the two Worcesters irrevocably. In Massachusetts, on 15 October 1684, the General Court granted that the Indian name of Quinsigamond be superseded by Worcester. The new name is believed to have been suggested because it was the English birthplace of some of the committee members or settlers. But some sources believe that the Massachusetts city was named for the Battle of Worcester, fought on the Worcester Plains in 1651 between the forces of Charles II and Cromwell.

Certainly, people emigrated from Worcestershire. St. Peter's Church in Droitwich memorializes Edward Winslow. That village near Worcester was the home of this *Mayflower* passenger who later became Governor of Massachusetts.

While many lament that the jumble of Tudor houses that once distinguished Worcester has been replaced by a modern jumble of commercial developments, multi-story car parks, and tortuous traffic patterns, there is nevertheless a great deal to justify making the 110-mile trip from London—or the trip across the Atlantic.

Wrentham

This small Suffolk village consists essentially of a long main street (the High Street) strung out along a main highway (the A12) which is lined with houses, inns, a town hall, and shops. The road was constructed in the old coaching days toward the end of the eighteenth century when Wrentham was a stage on the turnpike road from London to Ipswich to Yarmouth. The stage coaches may be gone, but the busy road maintains steady streams of traffic, particularly in the summertime.

Just three quarters of a mile away from the business and traffic of the village center, in a peaceful and pretty place, is the Parish Church of St. Nicholas. Its Perpendicular tower can be seen from a distance. Indeed, it was used as a signal tower to give warning of invaders in 1804 when England felt threatened by the Napoleonic invasion. While that scare came to nothing, the only real invasion remains that of holidaymakers rushing through Wrentham on their way to coastal resorts. Drivers who do manage to stop at this former coaching village can find recompense.

The Church of St. Nicholas, built on a Saxon site, is first referred to in Norman times. The exact date of building is not known, but it was extended in about 1260 with the addition of a chancel and expanded again in the fifteenth century when Wrentham prospered. At that time, the large Perpendicular tower was also built.

In 1853 the church was again extensively restored and the north aisle added. This newest part contains some of the oldest stained glass in the church, dating from the fifteenth century. One medieval picture shows the patron saint of the church, St. Nicholas, clasping his staff and mitre while a child is seated on his knee. Another bit of ancient glass depicts in exquisite detail an unknown saint.

The church has lovely touches everywhere. The tower doorway must be singled out for its beautiful carvings; and the nave,

252

for its clustered columns. Bearded heads are in the roofs, and two old memorial brasses are in the floor. Ele Bowet of 1400 is engraved in brass dressed in a long robe with bell-shaped sleeves, and Humphrye Brewster is decked out in Elizabethan armor. The Brewster family built Wrentham Hall in 1550 (now demolished) and made this one of the influential English strongholds of Puritanism.

Reading through the visitors' register in the church, one might come away with the impression that the only travelers to stop speeding and start looking come from Wrentham, Massachusetts. A connection is certain.

John Phillips, Rector of Wrentham in Suffolk went to Salem in 1638 where he stayed for about three years before returning to his former parish where he died in 1660. Indeed, many former residents of Wrentham in Suffolk left for the antipodean retreat from about 1660. In 1673, Wrentham in Massachusetts was created a town. It seems a form of retribution that the pattern has now been reversed: American visitors, full of reverence, return to Wrentham origins.

York

Just over nineteen hundred years ago, a Roman legion estab-
lished a fortress on the banks of the River Ouse for defense in
the north of England. They ousted a powerful British tribe, the
Brigantes, who had been settled in the area for centuries. Thus
began, in the year 71, Eboracum, the place of the yews—or York.
The city became one of the most important in the Roman
empire. The emperors Hadrian and Septimus Severus came to
Eboracum, and Constantine the Great was here proclaimed
Emperor in 306. Of the few visible traces that remain of the
Roman city, the most impressive is the Multangular Tower, the
west corner of the fortress. So much for the grandeur that was
Rome.

After the Roman legions withdrew near the start of the fifth
century, several successive waves occupied this important
center. It was the Saxon city of Eoforwic when King Edwin, a
convert to Christianity, built the first church in 627 on the site
of the present cathedral. Danish invaders took over in 876 and
made Jorvik (or York) their capital city. Old street names have
persisted from these times. The main shopping thoroughfare of
Coney Street was Conynge Strete, a name derived from the same
root as the German word *könig* or king; it was the king's
highway. Goodramgate is named after Guthrun, a Danish
chieftain. The Scandinavian suffix *gate*, meaning street, is
commonly found in many of York's street names: Petergate, St.
Saviourgate, Swinegate. Micklegate means Great Street, and
Stonegate may have been so named because it was stone paved.

Alas, poor Jorvik fell in 944 when Edmund conquered
Northumbria and made it part of the Anglo-Saxon kingdom.

Under William the Conqueror, the Normans built two
wooden castles atop earth mounds to subdue the rebellious
population. The mound which survives on the west bank of the
River Ouse is known as Baile Hill. The other is Clifford's Tower,
a thirteenth-century stone keep. The original Norman castle on

254

that mound was tragically destroyed in 1190 when the Jewish community of York sought sanctuary there. They set fire to the castle and died in the flames rather than be slaughtered by the bloodthirsty mob. A Jewish boycott of the city of York officially ended as recently as 1978, with a joint ecumenical service in Clifford's Tower.

Clifford's Tower

The medieval core has persisted through the centuries to give a unique and savory flavor to the modern city center. Much of the medieval thirteenth-century stone wall which surrounds the city offers good walking, but the most attractive part to walk on is between Bootham Bar and Monk Bar. A number of bars or gates are in good condition. Walmgate Bar is the only one in England with its barbican intact. Invaders passing through the narrow channel made by the outward extension and gateway of the barbican were subjected to a barrage of missiles. Micklegate Bar was entered by those approaching from the south, and on this gate were displayed severed heads of traitors impaled on spikes. Shakespeare has Queen Margaret exclaim in *Henry VI* of Richard, Duke of York:

> Off with his head, and set it on York gates;
> So York may overlook the town of York.

In fact, heads of executed rebels were exhibited as late as mid-eighteenth century. From an aesthetic point of view, however, Micklegate Bar is appealing enough to allow for suppression of associated horrors.

Within the walls of the city, a tangle of streets, churches, guildhalls, and historic buildings provide a medieval atmosphere. The Shambles is one of the best-preserved medieval streets in Europe. The name is a corruption of Fleshammels (from the Old English word, *shamel*, meaning slaughterhouse); it was the street of the butchers. Now it is the street of the tourists. Timber-framed houses with jettied stories almost touch across the road, making it possible for persons to lean out from upper stories on opposite sides of the street and shake hands. The Butchers' Hall is at Number 40. Number 35 was the home of Margaret, the Saint of York. Alleged to have hidden Jesuit priests, she was martyred in 1586 by being pressed to death under a heavy door piled high with stones. The house of Margaret Clitherow is now a chapel and a moving shrine to the canonized butcher's wife.

Whip-Ma-Whop-Ma-Gate is York's shortest street with the most photographed name plate. Its curious name may have derived from the whipping of petty criminals on the site.

Stonegate is a particularly fine street. It started life as the Via Praetoria, the great Roman Road from Londinium in the south which crossed the river and led to the headquarters of the legion. Mulberry Hall of 1434, probably the home of a rich merchant, indicates the prosperity of the medieval city. The so-called Stonegate Devil squats under the eaves of Number 33, a printer's house. The oldest dwelling house in York, the Twelfth Century House, stands behind Stonegate and is reached through a narrow passage.

The Shambles

Among other buildings of medieval note is the Merchant Adventurers' Hall, where wool, England's chief export, was brought to be weighed. No longer powerful overseas traders, the company is today more of a social body. But the Merchant Taylors' and Butchers' Guilds are still active.

Also active is the ongoing series of the York Cycle of Mystery Plays, a set of religious plays believed to date from the middle of the fourteenth century. The ruined St. Mary's Abbey, of about 1270, with great empty windows and richly carved arcading, is the effective backdrop for the triennial performance of the York mystery plays. And in the grounds of St. Mary's is the Yorkshire Museum.

Some eighteen churches remain of the fifty which once existed in York. But in medieval as well as in contemporary times, the city was dominated by the Minster. One of the majestic cathedrals of the world, York Minster sits on top of the remains of the Roman legionary fortress. Formerly an outlying missionary run by a group of clergy, it was the mother church of the whole of the North of England ministering to the people—hence its fitting title, *Minster*.

Completed in 1472, the cathedral took over two and a half centuries to build. The dimensions (over five hundred feet long, nearly two hundred fifty feet wide across the transepts, and over ninety feet high) make it the largest Gothic church in England.

The stained glass of York Minster is glorious. In the south transept is a circular window commemorating the marriage of the Lancastrian Henry VII and Elizabeth of York in 1486 which ended the Wars of the Roses. In the north transept is the largest thirteenth-century window in the world, over five feet wide and fifty feet tall; called Five Sisters, it is made up of five lancets of grayish-green or grisaille glass with colored geometric patterns. The great west window of 1338, the Heart of Yorkshire, has tracery in the form of a heart, and in the nave is a further priceless collection of fourteenth-century stained-glass windows.

The chapter house, a vast octagonal building with a conical roof, was a daring architectural feat when it was built in about 1300; external buttresses support the immensely heavy roof without a central pillar.

The cathedral is an endless source of fascination from the various tombs and monuments, astronomical clock, and fifteenth-century choir screen, to the crypt and Undercroft Museum with remains of the Roman Principia building and earlier churches. Steps to the central tower roof can be climbed for satisfying views of the city from this, the highest point of York.

Each century has left its marks and monuments on the city. Micklegate is essentially a Georgian street with many fine town houses belonging to the eighteenth century. But in busy Coney Street is the city's most ambitious Georgian building—Mansion House, where lord mayors live during their term of office.

View of York Minster from Goodramgate

Although York was largely bypassed in the nineteenth century by the Industrial Revolution, it did become a leading railway center under the instigation of George Hudson, whose house at Number 44 Monkgate is on the tourist itinerary.

And the beginning of this century saw the establishment of the Castle Museum, one of the most intriguing folk museums anywhere. The main building was formerly the Female Prison of 1780. Also part of the York Castle Museum is the former Debtor's Prison of 1705, with two projecting wings and central turret. The contents include a reconstructed Victorian cobblestone street with its apothecary, haberdashery, and other shop windows; the Edwardian Half Moon Court complete with shops and pub; craftsmen's workshops; and a cornmill which makes the stoneground flour which is sold by the bag.

The city of York petitioned the government in 1641 and 1648 for the establishment of a university. The granting of that request is the most significant event of the twentieth century. With the Elizabethan mansion of Heslington Hall as its nucleus, the University of York was finally opened in 1963.

Twentieth-century York, with its history still visible, is an accumulation of all the previous ages. And the American city of York, Maine—like so many other American cities—is indeed fortunate to be named for one of the most appealing cities in the world. The Maine name was bestowed in 1623 by Christoper Levett, who was born in 1576 in York, England. As Comden and Green might have sung their song if they had visited: York . . . York, it's a hell of a town!